D1490191

TECHNICALLY SPEAKING

A Guide for Communicating Complex Information

Jan D'Arcy

Battelle Press

Columbus • Richland

Library of Congress Cataloging-in-Publication Data

D'Arcy, Jan,
    Technically speaking : a guide for communicating complex
    information / Jan D'Arcy.
        p. cm.
    Includes bibliographical references and index.
    ISBN 1-57477-051-9 (softcover : alk. paper)
    1. Communication of technical information. 2. Business
presentations. 3. Public speaking. I. Title.
T10.5.D37   1998
601' .4—dc21   98-29188
                                                                                   CIP

Printed in the United States of America

Ordering Information
Battelle Press
505 King Avenue
Columbus OH 43201-2693 USA
Phone: 1-800-451-3543 or 614-424-6393
Fax: 614-424-3819
E-mail: press@battelle.org
Website: www.battelle.org/bookstore

TECHNICALLY
SPEAKING

WHAT OTHERS HAVE SAID

"Jan has worked 25 years as a communications specialist and professional actress, a synergistic mind-meld that is a constant winner. Read this book. Try out a few gems and be surprised by the reaction. It's a solid investment in your future."

Frank Ogden, Dr. Tomorrow

"The text of *Technically Speaking* is a practical and basic guide in assisting technical professionals to communicate effectively. This capability is becoming a more important skill with the rapid change in technology and the realization that technology understanding is critical both in the business world and in the community. This text covers most if not all the issues faced by presenters and shows how to overcome the common barriers to effective presentation."

William E. Coyne, Senior Vice President, Research and Development, 3M

"Jan D'Arcy's approach is results oriented. We used presentation techniques outlined in this book to help us win a multimillion dollar contract."

Vivianne Larkin, Vice President, Environmental Services, URS Greiner

"The biotechnology industry has remarkable benefits to offer the public, but these can only be communicated free of jargon. D'Arcy describes well the art of delivering the right message to the right audience."

Carl B. Feldbaum, President, Biotechnology Industry Organization

"Technical types have two choices: Read this book or lose your audience. Yes, this is an intelligence test."

Guy Kawasaki, CEO, garage.com

"Chapter Eleven is pure gold."

Patrick D. Moneymaker, Rear Admiral, United States Navy

CONTENTS

PREFACE

This book is a major expansion and updated edition of *Technically Speaking* which was originally published in 1992. In the intervening years, I have used the book in providing consultation and training to scientists, engineers, and high-tech personnel. Their feedback confirmed the usefulness of my approach.

The most significant change I have seen in the past few years is the growing expectation that a presenter use sophisticated electronic technology to present data in a meaningful way. I have therefore created a process for the development and design of "visuals from low tech to high tech." I have also included updated information on room and desktop videoconferencing and a chapter on international presentations. I have continued to interview model communicators and have incorporated new insights and secrets of success into this book.

There are many people whom I wish to thank for contributions to this book. I have nothing but praise for Bob Schultz, whose valuable expertise and intensive editorial efforts guided me through this project. Lara Chapman provided first-hand knowledge on international communications, Charles Lord contributed helpful information on researching methods, and Brian Painley supplied real-life examples of preparing engineers for cutting-edge electronic presentations and practical advice for avoiding the minefields. I am particularly grateful to Michael Treadwell and Carmen Smith for their assistance and encouragement. Many thanks to Joseph Sheldrick, my publisher at Battelle Press, who calmly kept me on track, and to Bea Weaver, Amy Householder, and Sharon Manwering, who were helpful in designing and laying out the text and graphics for the book.

I want to acknowledge my supportive children (Lisa, Paul, Colleen, Shane, and Tyler) who understood my time and commitment to the book. A special thanks to Shane who helps design and create the animated graphics for my laptop and Web presentations. He provided many insights that I have included in the chapter on visuals. Thanks to J. Wendell Forbes, Michael Stefan, James Lambert, and CH2MHill for their graphic contributions. I am especially indebted to the companies that have hired me and the thousands of individual clients and workshop participants from whom I have learned so much. And finally, I wish to thank all the model communicators who shared their ideas and inspired me to write this book.

I dedicate this book to my parents who gave me a lifetime of values and unconditional love, and who supported me in everything I ever attempted.

INTRODUCTION

This book is for scientific and technical professionals. It provides a system that can make oral communication of complex information *understandable* and *useful*. It is a practical guide that will enable the reader to make presentations to both fellow professionals and lay audiences in an efficient, effective, and clear manner.

In a world where technical knowledge doubles every 18 months, technical professionals increasingly find themselves in unaccustomed roles that demand new communication competencies. Companies battling to gain a competitive advantage and organizations fighting for funding must be able to translate the latest technological and scientific findings for the benefit of customers, colleagues, and key decision makers.

Scientists and engineers recognize that, although excellence in initial research and design is extremely important, marketing and selling to management and consumers is equally important. The ability to make a compelling presentation can be the deciding factor in determining whether a contract is awarded or a funding request is granted.

As presenters of technical information, you face a unique set of challenges. You must:

• Organize your material in strict sequential order

• Present it with precise clarity

- Use specific terminology

- Condense an overwhelming amount of data

- Be exceptionally competent and adapt your subject matter to audiences with different levels of knowledge

- Acquire a proficiency and comfort level in using a variety of technological presentation media

- Decode and provide creative insights into subject matter that is rigidly defined and may be difficult to see, touch, measure, or imagine.

Technically Speaking: A Guide for Communicating Complex Information is designed to lead the reader through the process of researching, organizing, and presenting such complex information, meeting all of the challenges listed above. It describes a ten-step approach that will save considerable time as you prepare speeches, briefings, and presentations.

Each chapter begins with an overview and concludes with an outline of key concepts. Checklists, worksheets, and exercises are included. There are suggestions for criteria for audiovisual materials, a discussion of electronic presentations, considerations when communicating to international audiences, and detailed guidelines for analyzing audiences.

In today's information-overloaded environment, it can be extremely difficult to capture and retain an audience's attention. When the stakes are high and the audience challenging, is it any wonder that performance anxiety rears its ugly head? This book emphasizes techniques to help you manage those anxieties that accompany public speaking. Such fears beset all speakers, but are particularly unsettling to people unaccustomed to taking center stage. Throughout the book you will learn how humor, metaphors, and anecdotes can be used to clarify factual data, transform it into useful knowledge, and perhaps even transmit wisdom.

This book is a result of in-the-field research and my twenty-seven years as a communication specialist conducting presentation skills seminars for thousands of people in colleges and universities, government agencies, and financial, legal, and medical institutions. I began to focus on the communication needs of engineering, high-tech, and bio-tech clients, and resolved to develop methods particularly appropriate for them. While preparing to write this book, I conducted more than three hundred interviews with personnel at engineering, high-tech companies, and scientific companies. I asked heads of corporate departments of research and development, education and training, public relations and communications, and sales and marketing to name model communicators within their organizations. I then interviewed these "star performers."

These model communicators were acutely aware of the special difficulties of communication in their fields and were eager to share their speaking experiences and expertise. The information and insight gleaned from these interviews are an integral part of this book.

Technically Speaking: A Guide for Communicating Complex Information can be used two ways: as a how-to instruction guide and as a reference tool. The novice speaker can learn the basic skills of oral presentation. The experienced speaker can use the book to refine communication skills and to update presentation techniques.

The physicist Richard Feynman believed that "making things as plain as day can also make them as sublime as the night sky." He demonstrated that the beauty of science is best conveyed not by wrapping it in mysticism or poetry, but by describing it simply, passionately, sometimes comically, and always clearly. With the help of this book, you will be able to do the same.

Part I

Becoming Comfortable, Confident, and in Control

Chapter 1

BECOMING A MODEL COMMUNICATOR

"We are what and where we are because we have first imagined it."
—Donald Curtis

OVERVIEW

Effective communication is a skill that you can learn. All you need is desire and a guide. This chapter begins your plan of action.

This time, you tell yourself you're going to give the perfect presentation. Given enough lead time, preparation, and self-confidence, you know you have the ability to present the kind of speech that will cause your audience to sit up and take notice.

You picture the surprised looks from the company's senior officer and the rapt attention of those seated in the back of the room as you make one insightful point after another. Heads nod in agreement as your visuals clarify the complex information. Hands shoot up to get your attention without your asking for questions. It's a lively give-and-take session. Applause follows and you feel a glow of genuine satisfaction as people crowd around you offering praise and requests for further information.

Ah, if it could only be like that!

Reality sinks in. You realize that you'll probably end up giving another slapdash report with a few basic visuals thrown together at the last minute—due in large part to deadline pressure and lack of preparation time. You're probably dealing with soft information that needs verification. You're uncertain about your audience and what they want to hear. Maybe you're overwhelmed by other work responsibilities that demand your attention. Or you're busy attending to critical events in your personal life.

Frustrated, you think that if only you could take all those ideas running around in the corridors of your mind, isolate them, line them up in formation, and skillfully communicate them to your colleagues and the public, everyone would realize how brilliant and perceptive you are. And a myriad of problems would be solved.

Suddenly, your anticipation of getting the sale, convincing the review board that your project is on track, persuading a group of investors that your product or service will be profitable for them, or having your research validated by your peers changes into the dreaded possibility of making a fool of yourself. The mental image you had of giving the perfect presentation is replaced with Murphy's Law: Anything that can go wrong will go wrong. And somehow you have equated the failure of the speech with failure as a person.

If you've ever felt like this, you're not alone. The fear of public speaking is universal. No matter how experienced or how lofty the position, virtually every person faced with public speaking experiences some form of stage fright. And there are countless others who avoid speaking at all costs. Effective communication is rarely an accident. Designing and delivering a successful presentation is a skill. The good news is that effective communication, like any skill, can be learned.

I am cautious about agreeing with someone who says, "Well, I certainly learned from my mistakes that time!" I hear the catch in the voice, sense the tension in the muscles, and see the tightness in the jaw. Because, although it is possible to learn from one's mistakes, I also know that a technicolor imprint of those mistakes is ingrained into the memory forever. When that happens, it usually takes teeth-gritting willpower to block out negative memories and face an audience again.

You can learn to swim and stay afloat without any coaching; it is also possible to muddle through reports and sales presentations alone. Eventually you may succeed. In swimming, a coach can teach us how to stop thrashing about. We learn to breathe rhythmically and refine our strokes. We learn to merge with the water and not fight against it. And with every lap, we cut down the time it takes to swim from one point to another.

In speaking, we can also learn the refinements and the subtleties of getting our ideas across to our audience. We can learn to use our fears and anxieties to enhance our delivery. We can use techniques that will help save preparation time. Each positive experience will encourage us to look forward to opportunities for sharing information. Speaking in front of a group can bring satisfaction and pride. It can be a very enjoyable experience.

But the learning process begins with a conscious decision. Just as you commit yourself to excel in your profession, you must also commit yourself to excel as a public speaker and do what it takes to succeed. Let's get started!

Chapter 2

COMMUNICATING COMPLEX TECHNICAL INFORMATION

*"Everything is simpler than you think and at the
same time more complex than you imagine."*
—Johann Wolfgang von Goethe

OVERVIEW

*The information explosion does not necessarily mean there has been a knowledge
explosion. We are data rich and understanding poor. Your audience needs to understand your ideas before they can accept them. This chapter encourages you to look at
the complex data you are presenting and find ways to tame, tailor, and illuminate that
information so you are transmitting useful knowledge that will be accepted by your
audience.*

The mind is only capable of absorbing so much information before it shuts
down. A Rear Admiral, a Naval aviator, told me that the pilots of multimission
aircraft can be so inundated with computerized information that there are
moments during task saturation when they can be dysfunctional (sometimes
called a "helmet fire"). The Navy installed a switch on the HUD (heads-up
display) of the F-18 that toggles from normal to de-cluttered mode. This
enables the pilot to manage the amount of data he receives. This is a first
step in limiting the input to keep the pilot at peak performance.

Audiences also become anesthetized when verbally and visually overloaded.
A confused mind will say, "No!" The presenter who extracts only the essential
information from the vast amounts of available complex data and communicates
this in an understandable and useful way will elicit applause as well as sighs of
relief from the audience.

All of us must invest an enormous amount of energy in absorbing and processing information. A friend of mine was struggling to learn Akkadian, an ancient dead language. Making sense of the writing was an extremely frustrating task. Her teacher finally admitted that the only way to decipher an Akkadian tablet was to know in advance what it said. He explained that you need to know the subject, context, and jargon so you have a frame of reference and can choose the correct reading of a graphic symbol. When you are caught up in your ideas, it is easy to forget that other people don't have the same experience and familiarity with your technical discipline. Just as my friend needed prior information to learn Akkadian, you will need to provide basic references for your audiences.

Never overestimate your audience's knowledge base; at the same time, never underestimate their intelligence. A computer programmer told me that she had not been working in her field for eighteen months. When she went out for job interviews, she not only didn't know the workings of current software programs, but didn't even understand the acronyms referring to them. She was depressed because she couldn't demonstrate how intelligent she was to the interviewers. They had already predetermined that she was not qualified because she didn't belong to the current "information club." She became competitive only after researching the market and taking time to update her skills. Assume your audience is smart, but not up to date in your discipline.

Model communicators make information simple and easy to understand without watering down their ideas. Albert Einstein said, "Everything should be made as simple as possible, but not simpler."

The most interesting discovery that I made in the course of my interviews is that model communicators illuminate and give insight; they don't dilute scientific and technical information. They don't talk down to their audiences; rather they feel that scientific concepts should be available to everyone. They empower their listeners. They "translate" complex concepts in a way that adds to a person's knowledge. Mike Sundell, Vice President and General Manager for Basic Coatings (an Iowa company that produces wood and metal finishes), said, "I think in terms of bringing technology up to the level of my audience. A lot of audiences are intelligent and sophisticated; they simply haven't been exposed to this specific technology."

Publishing consultant J. Wendell Forbes remarked:

> No matter at what level of management we operate, and no matter what our job is, the challenge to all of us is to understand that whatever we are doing, the ultimate goal is wisdom. Even if we are on a production line, our data is what is delivered to us and our wisdom is what we should deliver to the next person. None of us operate exclusively at the wisdom level. We all start with our own equivalent of data and strive to end up with the ultimate satisfaction and reward that we here characterize as wisdom. One person's wisdom is another person's data.

Strive to pass on to your audiences useful knowledge and wisdom that will help them reach their goals.

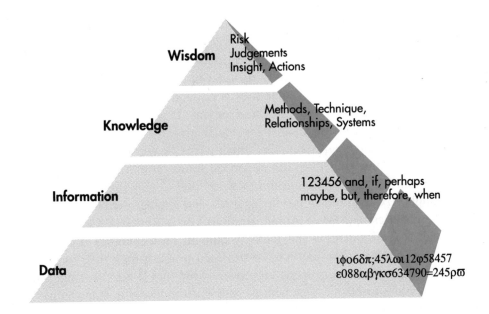

Wisdom Triangle
Courtesy of J. Wendell Forbes

RECOGNIZING AND RESPONDING TO MULTILEVELS OF KNOWLEDGE

How can you translate complex material so that your audience receives the information easily and can incorporate that knowledge into their experience base? Every individual in an audience has a different world view, and there may be as many levels of knowledge as there are people. An audience analysis will help you understand how to present your material.

Let's say you are going to address three different groups. The first audience may read *Reader's Digest* and *USA Today*. This audience will want and need to know very simplified versions of technical information. A second audience might read *Scientific American, Omni,* or *Byte* and attend computer user or environmental focus groups. This level would also include people who give technical business presentations. This audience will feel comfortable with some technical information. Another audience may attend technical conferences that present in-depth information, statistics, product details, demonstrations, references, samples, plots, schematics, and drawings. This audience is interested in the hows and whys. Their reading material might include publications such as *American Institute of Aeronautics and Astronautics Journal, Journal of Applied Physics,* and *IEEE Transactions on Geoscience and Remote Sensing.* (IEEE is the Institute of Electrical and Electronics Engineers.)

A researcher should be able to convince a budget review panel that her project is valid and profitable for the company as easily as she would address her peers to describe the details of the process she followed. And she should also be able to adapt that same technical information to a general audience.

The most difficult presentations to prepare are those to audiences who have widely varying levels of knowledge. Begin your presentation by covering some basic information and terms for the people in your audience who have little knowledge of your subject. Periodically mention in-depth points that will keep the interest of the more knowledgeable members of the audience. Always give definitions of technical terms and explain acronyms. If it is necessary to describe a complex concept, give examples that include reference points familiar to everyone. Reiterate your main points in simple terms as you go along. End your speech by summarizing your message in general terms so that your entire audience can reach the same conclusion that you do.

For example, I was asked to coach scientists and engineers from NASA (National Aeronautics and Space Administration) Lewis Research Center who were to speak at their Business and Industry Summit on technology transfer from the space station. These presenters were used to explaining their work in technical settings with homogeneous audiences. Now they were challenged to adapt their message to business audiences with diverse levels of knowledge. They were addressing groups made up of nontechnical executives interested in profits and partnerships and of technical R&D (research and development) specialists who wanted to evaluate product design and development.

The objective of one of the speakers was to convince the business people to incorporate a valve design from the space station into their products. He began by giving a general overview of the valve's design and its use in space. Next, he gave an example of the valve's versatility by demonstrating how this design had been adapted to medicine. He displayed a titanium heart valve that can success-fully be used as a substitute for a heart transplant. He pointed out some sophisti-cated technical details about workings of the valve and carefully defined his terminology and acronyms. Then he explained how the design of the heart valve could be applied to improving the design of valves in other products. He ended his speech talking about the cost and performance benefits rather than dwelling on all the complex details that were far too technical for the key decision makers in this business audience.

How Much Detail?

David L. Harten reported in *Reader's Digest's* "Campus Comedy" that a liberal arts student scrawled the graffiti, "Love makes the world go round." Under-neath, a physics student added, "With a little help from intrinsic angular momen-tum."[1] Are you putting off your customers by going into too much detail? Is that why they decided to buy from your competitor? What did your customer *really* want to know? Scientists and engineers who think in precise terms often com-municate in the same manner, but you can alienate an audience by talking over their heads. Avoid exactness if the audience's knowledge level dictates a more general approach. The perceptive expert will seek ways to diminish the gap between his knowledge and that of the audience. If the listener begins to feel ignorant, he will resent the presenter and his purpose.

Give the general idea or major findings before you give the details, the function, and the principles. Alert your audience that you are going to build a skyscraper, not a log cabin. Then you can describe it and show them how step by step. If you are describing characteristics, help your audience by zeroing in on the specifics. Decide how much detail your audience needs to know to do the job, make the decision, advise others, or achieve some clear purpose. While one audience may feel buried in particulars, a technical group of your peers may think that four days of in-depth talks and a four-inch notebook merely skim the surface.

Decoding Tech-Speak Into Concrete Images

There is nothing frivolous about depicting complex scholarly subjects in understandable concrete images. Choose images that are familiar to your audience. A scientist from Dupont said, "Spider silk is the toughest material known. It is also very elastic. On an equal weight basis, it is stronger than steel. It has been suggested that a single strand of spider silk, thick as a pencil, could stop a 747 in flight." Illustrate the application of a theory or demonstrate your conclusion with words that engage the senses of touch, sight, smell, or taste.

Robert M. Price, President of PSD, Inc., clearly illustrates the complexity of parallel processing with this simple comparison:

> Without using parallelism, you could simply hire one person with one lawnmower to mow your lawn, but you have a large lawn and it will take one person four hours. To shorten the time, you could contract with four people to do the job: one person for each side of your house. The control is simple: A mows the front, B mows the back, C mows the left side, D mows the right side. In this case, you have used parallel processing to reduce the time needed to mow your lawn from four hours to one.

> Let's push this method and hire 240 people to mow the lawn. Can you expect that the job will be done in one minute? Not exactly. The problem is that you must spend a lot of time contacting each of these people and telling them what to do so that they aren't running over each other with their lawnmowers. With 240 workers, the simple job of mowing the lawn becomes a major task of control.

> In general-purpose computers, we know how to manage modest levels of parallelism. But we don't know how to manage large numbers of parallel processors effectively.

Avoid "fat" words or abstract words. Quality, change, and productivity are fat words. For example, if you say CADD (computer-aided design and drafting) is a high-quality software program, are you saying that it is faster (how fast?), is more interactive, or has more pixels per inch?

Defining Your Acronyms and Buzz Words

One high-tech trainer told me, "We get so buried in our technical language that we forget that others don't have any idea what those acronyms are referring to." An audience may miss the whole object of your presentation because you assume they understand terms basic to your own profession. A nontechnical audience may hear you say ATM and think automatic teller machine instead of asynchronous transfer mode.

An overabundance of acronyms is especially prevalent in government and high-tech circles. TechWeb's Technology Encyclopedia (www.techweb.com/encyclopedia) lists over 10,000 acronyms.

Use all of the words in an unfamiliar acronym more than once so that your audience can remember exactly what it means. It might be advisable to include a cue sheet or glossary among your handouts.

If the words used to form the acronym are a meaningless abstraction, then the resulting acronym can be totally vague. For example, everyone knows lasers are something we encounter daily in the barcode scanner in the supermarket. Many people know astronauts communicate with earth through laser beams. Lasers are used heavily in medicine; they can cut and cauterize certain tissues and bore holes in the human skull. However, a recent poll asked over 2,000 adults if lasers worked by focusing sound waves. Thirty-six percent correctly answered that it was false, but 29 percent thought it was true, and 35 percent didn't know. Even if they knew the acronym stands for *Light Amplification by Stimulated Emission of Radiation*, they might need a further definition of the terms to understand how lasers work.

People are intimidated by jargon but often won't ask for an explanation. Last year I attended a seminar about doing business on the Web. The presenter did an excellent job of showing how to incorporate the Internet into business strategy. However, he kept mentioning "cookies" during the first two hours of the presentation. For example, he stated that "cookies are the perfect tool to make a portrait of your customers." Despite the context, it was difficult to figure out exactly what he meant, although "cookies" were obviously important.

During the break, I asked a number of attendees if they were using cookies. "Chocolate chip? Or peanut butter?" one person responded with a laugh. "Haven't a clue what he's talking about," said a puzzled young man. "Didn't want to reveal my ignorance by asking," responded another. I went up to the presenter and suggested he explain the term.

When we reconvened, the presenter started off the rest of the seminar by explaining that cookies were one of the newest methods of electronic tracking and information gathering. He described cookies as files stored in the visitor's computer that hold some information about the computer, record the pages on the site visited, and stores other information the visitor provides when visiting the Web site. The Web site assigns an identifying number to the user, determines the Internet service provider, and documents the path (clickstream) from origin to destination, revealing browsing interests. If the visitor returns to the Web site, this file is—if the visitor permits—sent to the Web site and any new information is added. By examining these cookie files, the site can gain information about its visitors and by tracking their path through the Web site, can identify and respond to their preferences. A perfect marketing tool, indeed!

Meaningful Definitions

> *"An inventor is an engineer who doesn't take his work seriously."*
> —Charles Kettering

Definitions only work when your audience is familiar with references in the definition. What is a Xanadu? If I said that a Xanadu has the tail of an elephant and the neck of a giraffe, you could start to imagine the creature. But if I said that it also had the body of a tripozip, your mind would have a difficult time searching for an association.

George A. Keyworth II, the director of research for the Hudson Institute, added to his audience's knowledge with a thorough definition:

> There's one word I want to use that sounds technical, but needn't be. The word is digitization, but think of it as a computer-age version of Morse code, the old Western Union language that had only dots and dashes. Like Morse code digitization consists of only two words. Those two words are the means by which information is made so simple that it can be treated by the computer as nothing more than a series of ones and zeroes to be added and subtracted. But what makes digitization significant is that virtually any kind of information—and by that I mean words, numbers, voice, music, photographs, or movies—can be converted into those streams of ones and zeroes.[2]

A definition can be made more explicit by comparing it with something familiar. For example, a DVD (Digital Versatile Disc) is about five inches in diameter and stores binary data in video and audio format in microscopic pits on the surface. This two-sided disc has two "layers" on each side and provides more than 25 times more data-storage capacity than a CD or a CD-ROM. A DVD provides higher-quality sound comparable to a good movie theater. The image quality and resolution is twice that of your VHS tapes. Two hours of video or the length of a feature film can be stored on a single layer.

What Is It Like? Similes, Metaphors, and Analogies

Your audience will show discomfort, annoyance, and frustration if they are unable to understand your presentation. No one likes to think that he or she may lack the intelligence to grasp information. Many scientific subjects are hard to describe; they can be difficult to see, touch, measure, or imagine. A presenter should seek to find ways to illuminate a concept in known terms with the least amount of distortion. Aristotle said that people remember information better if it is:

- closely associated with something familiar,

- sequential, or

- contrasts with something they already know.

Comparisons and contrasts are two of the best ways to translate your information clearly to your audience. Similes, metaphors, and analogies are comparisons that often can lead to amazing insights. By using these devices, you can enliven the dullest data and help people discover they know more than they thought they did.

Simile

Makes a direct comparison between two dissimilar objects and always uses connective words or phrases such as "like," "as," or "as if." For example, research is *like* a treasure hunt. Leonard Pitts, a *Miami Herald* columnist, recalled his incompetence on the dance floor: "I stood there *like* a totem pole in a body cast." Fresh, unique similes enliven your language.

Metaphor

Assumes an identity between two things. A metaphor is a condensed simile. For example, research *is* a treasure hunt. Metaphors are useful descriptions that add color and express feelings: "The software marketplace *is* a jungle." You are substituting an image for an idea. The names of athletic teams are often metaphorical: Cowboys, Seahawks, Pirates.

Analogy

An expansion of a simile or metaphor. It uses the similarity of attributes, uses, or circumstances between two objects or concepts to explain an unfamiliar object or a concept. Whereas similes and metaphors use vivid language to get an audience's attention, the analogy facilitates understanding by explaining the complex in simple, everyday terms. In the movie, *My Best Friend's Wedding*, Julia Roberts uses an analogy to explain to her rival that the man they both love doesn't necessarily want perfection but wants something comfortable and familiar. "As hard as it is to believe," Julia Roberts' character says, "some people prefer Jello over crème brûlée."

Mike Sundell of Basic Coatings, speaking before a nontechnical audience, explained the purpose of adding catalysts and cross-linkers to polymer resins as follows:

> To make products dry hard enough for floors, it has always been necessary to add catalysts or cross-linkers. Think of polymer resins in the solution as individual balls. If you tried to walk among the balls, you'd fall through. But if you tied string between them, you'd form a net that would hold your weight. That's what cross-linkers do: They tie the polymers together.

Mike notes that the analogies he uses to make a point early in his sales presentation are sometimes shaped by regional differences in his audiences.

> An audience in the heart of New York City is a natural Doubting Thomas with a show-me attitude. They listen differently to you than the people do in the central corn country of Iowa, where they don't lock their houses or cars at night. I check to see how my audience is listening to me and how they react. Then I know how to tailor the rest of my presentation and what type of analogies they will respond to.

University of California Irvine biophysicist Bruce Tromberg is studying the "flight of photons" or how light particles travel through different kinds of tissue. Tromberg compares the photon flight paths to drivers on the freeway: You may switch lanes a lot (akin to light scattering) or make pit stops (akin to being absorbed). After Tromberg's team members shoot light into tissue, they can tell exactly how many times photons changed lanes and how many made pit stops. Their work may lead to doctors using portable laser detectors to learn whether breast tumors are cancerous, diagnose elusive cervical cancers, and map brain tumors—all without surgery.

You may agree that analogies help clarify points and are an excellent way to increase understanding, but how do you make up an analogy?

- Think about the situation or idea you want to describe.

- Think about what you want your audience to feel, think, and do.

- Review your audience's background, particular interests, and knowledge base (Analyzing Your Audience, Chapter 6).

- Find several things in your *unfamiliar* idea, process, sequence, action, or object that have similarities to a *familiar* idea, process, sequence, action, or object. What are the interrelationships?

- Choose familiar objects, actions, places, people, myths, sports, or experiences that do not require further interpretation. Your audience should be able to make an immediate connection and comparison. Note that the impact from the above analogies came from using ordinary, familiar objects.

- Make your analogy short, simple, and fairly general so that your audience can now view the idea in a new way.

- The essential task is to identify analogies that will be meaningful to your specific audience. Carol Bartz, Chairman and CEO of Autodesk, Inc., recognized that her technically literate audience at the Business Week Conference on the Digital Economy was familiar with the Web. She used changes in the exchange of information before and after the Web to explain parallel changes in partnership styles in the digital age.

> The Web...is open and accessible.... In the past you had a better chance to control what your partner knew about your company. Like opening and closing a faucet, you had the option to manage information flow to a few people or a few organizations at a time. Now, it's more like a floodgate—stuck in the open position.... In the past, you could shape a message and deliver it. Now you need to keep shaping and reshaping information in play as it evolves and changes. It's like changing a tire while the car is moving.... Openness means that spin control is ongoing and carried out in real time.

One of my clients, Mike Carson, an engineer from EBASCO (Electric Bond and Share Company), was summoned to be an expert witness in a court case. He was testifying about the condition of the utility system, including the poles and other equipment, and the extent of depreciation in the utility district. It costs $2,481 to install a new utility pole. According to the tax laws, a pole has no "book value" after thirty-five years, yet in the dry climate of northern California, poles may last twice as long as this. Therefore, the question was, What is the value of the average utility pole? The complex information would have overwhelmed a jury of ordinary citizens.

I asked Mike to think of something that the jury could relate to that would compare with the depreciation of telephone poles, as described above. We started to work out an analogy with a house depreciation and replacement cost. However, we were concerned that some of the members of the jury might not be homeowners and that they wouldn't recognize the relationship. We abandoned that analogy, and Mike suggested that almost everyone has bought and sold a car. Book value and depreciation rates would be familiar to them. When the attorney asked Mike to explain, he said, "It is similar to the depreciation of a car," and proceeded to make a clear comparison.

Because science and technology are constantly changing, new ideas and information can sometimes be explained only in terms of what we already know. Similes, metaphors, and analogies are useful tools to clarify communications.

FAILURE TO COMMUNICATE

If the response you receive from your audience indicates that your material is hard to follow, ask yourself why it is obscure. Do you avoid analyzing and interpreting because you fear making inappropriate value judgments? Are you concerned that your views will conflict with those of a superior? Is using tech-speak your way of conforming and insulating yourself from others, especially outsiders? Are you overly conscious of criticism and therefore qualify every statement? Of course, one final possibility (and one that I hope isn't true) is that you haven't done your homework and have nothing insightful to say about your topic.

Your audience will appreciate your efforts to decode and interpret complex information. A judge who was presiding over a difficult trial said that the defense lawyer presented layers and layers of documentation that included federal antitrust litigation, laws from two state jurisdictions, environmental issues, and contract and employee rights litigations. She noticed that the jurors' eyes glazed over in confusion and boredom as they attempted to follow the bewildering onslaught of facts, statistics, and details over a six-week period.

The defense attorney gave a long and tedious closing argument, reciting multiple facts from the case and complex legal theories. After making a simple presentation in his closing argument, the plaintiff's attorney advanced toward the jury. "What we have here," he stated, "is a classic case of a fox getting caught in the henhouse. Now all you have to do is decide how much the fox has to pay," and he sat down. The jury members relaxed with a sigh of relief. They understood this language. The plaintiff was awarded the largest amount of damages in a state court in the history of Washington.

Some people can make the trivial complex. What is simple is always a matter of subjective assessment. What is understood becomes simple. Programming your VCR isn't simple if you're not sure how to do it. However, piloting a helicopter might be simple if you've been thoroughly trained.

An audience that doesn't understand the data you are presenting will remember little. Nor will they be persuaded to buy your product, fund your research, accept your bid, or be influenced by your ideas. Use distinctive language that does not obscure meaning. Understanding is a prerequisite to acceptance. Take responsibility for the audience's ability to understand your topic.

KEY IDEAS

- Empower your audiences by illuminating complex information.

- Strive for clarity and use concrete words and examples.

- Associate new information with the familiar so that your audience can see relationships.

- Use stories and analogies; they will be remembered longer than dry facts and statistics.

- Never underestimate the intelligence of your audience.

Notes

1. David L. Harten, in "Campus Comedy" column. Reprinted with permission from the September 1986 *Reader's Digest*. Copyright © 1986 by The Reader's Digest Assn., Inc.
2. George A. Keyworth II, "Goodbye Central: Telecommunications and Computing in the 1990's," *Vital Speeches of the Day* (April 1, 1990).

CHAPTER 3

MAKING FEARS AND ANXIETIES WORK FOR YOU

"Of all the liars in the world, the worst are our own fears."
—Rudyard Kipling

OVERVIEW

You may be a brilliant engineer, scientist, or technologist, but if fears and anxieties adversely affect your ability to communicate, you will not be able to inform, influence, or persuade others. In this chapter, you will learn to specifically identify your fears and then develop a strategy to deal with them. If you minimize the perceived threat to your self-concept, you will find that your physical reactions will diminish.

Recently, I received a telephone call from the chief executive officer of a software company. He wanted to discuss his fear of public speaking. He said he was a "hard core" case and panicked at the very thought of speaking in front of a group.

When I met this gentleman, he confided that even though he had avoided public speaking, his company had grown and he was financially successful, because his product was in such demand. Now he was asked to speak about his software in the international market, and his buyers in Europe didn't want a substitute. In addition, he said he had some innovative ideas about civic causes that he wanted to address. He wanted to gain community support. Could I possibly help?

This gentleman was articulate, dynamic, and exceptionally intelligent. Yet I wasn't surprised by his admission. Over the years I find that no matter how lofty the position, the social status, or material success, somewhere deep inside every one of us is the feeling that we just aren't good enough, that somehow we will make fools of ourselves in public.

I accepted the offer to work with the software executive, but he told me he would have to postpone our first class. He was flying his plane to Montana for a week of survival training with an Alaska bush pilot. "But I would be scared to death to fly under the conditions they do," I said. "What's to be scared of?" he asked in puzzlement. A threat of real bodily harm was not as frightening as the perceived threat of an audience judging him and his ideas. There are millions of individuals like this CEO who avoid speaking at all costs.

Virtually every speaker, every performer experiences some form of stage fright. Quarterbacks for the Super Bowl related their overwhelming anxiety as they prepared to take center stage in the game's biggest showcase. "The Super Bowl was absolutely the worst, the worst experience for me," commented Terry Bradshaw of the Pittsburgh Steelers. "It's nerve wracking and all you have is time to think about all the negativity." Joe Theisman of the San Francisco '49ers said, "For a second, you are paralyzed with fear." His final thought as he ran onto the field for pregame instructions was, "Don't trip!" Rex Harrison (who played Professor Higgins in *My Fair Lady*) said, "There isn't a performer with an ounce of talent that is completely relaxed in front of an audience." Take pleasure in the fact that only fools know no anxieties.

The path to success begins by *acknowledging* and *accepting* your fears. To deny you have fears is counterproductive and could lead to disastrous results. Everyone experiences fear to some degree. It's a normal, natural emotion that cannot, and should not, be suppressed. Recent evidence suggests that the intensity of the fear emotion can be attributed to genetics. Other studies report that the initial programming of a humiliating or embarrassing situation magnifies the fear. Gradually introducing elements of the speech situation can be effective. Researchers have also shown that virtual reality simulation can reduce anxiety and help speakers control their fears. There is no one cause or one therapy. We can, however, learn how to make fear help, rather than hinder, our performance.

ANALYZING YOUR FEARS

Dave, an engineer friend, called me with a problem. "Jan, I have to make a presentation next week," he said, "and I'm so anxious about doing well that I haven't been able to sleep for the past week. Can you help me?" His voice sounded as if he had been condemned to stand in front of a firing squad, and he wanted me to make it less painful.

"Can you tell me exactly what you are afraid of?" I asked. "I'll probably make a fool of myself," Dave answered. "What if I can't remember all the points? What if I can't answer all the questions? What if my speech goes overtime? I've wasted so much time worrying that I only have a few days left to prepare." Half-jokingly, he added, "Isn't there some scientific way to get rid of these fears?"

I told Dave that I would help him, and no, there wasn't any scientific way to eliminate his fears. Besides, to completely do so would make him an ineffective speaker. "Fear is nature's way of helping you be alert, sharp, and up to doing your best—like the pregame tension of an athlete."

Most of us, like Dave, can drive ourselves to distraction and waste untold energy by dwelling on "what if's." Fear is caused by a perception of a threat. Some fears are real, but most are imaginary. As Mark Twain observed, "I have had a great many troubles, but most of them never happened." A scientist takes precautionary measures in the laboratory to avoid accidents; there are also measures that one can take to ensure success on a public platform.

"Small amounts of fear alert you," explains Philip Gold, Chief of the Clinical Neuroendocrinology Branch of the National Institute of Mental Health. "We think better, we move faster, and we adapt better. But too much stress tips the scales the wrong way. Then we can't remember, we can't adapt to a situation. We make wrong choices. We feel demoralized and lose confidence in ourselves. It can be catastrophic."

One way to keep from being overwhelmed by our fears is to analyze what we're afraid of—such as forgetting our lines or not being able to answer everyone's questions. When we analyze those concerns, we realize that solutions can usually be found, whether by thorough preparation or by developing effective techniques.

I've noticed my scientific and technical clients are prone to stage fright because they tend to have an aversion to any kind of "performing." They are usually reactive in their communications, even in their social lives. Separating themselves from the group—which is also frowned upon in some cultures—and taking center stage is unpleasant and strange.

They are more comfortable observing and reacting, which is why they excel in the question and answer period, rather than in making a prepared presentation. We create our own insecurities and fears. Unfortunately, we make decisions based on the fantasies our imaginations conjure up for us, or we make decisions based on an attitude shaped by something that happened years ago—or what we *think* happened.

When I taught communications in the Management Program at the University of Washington, I sent a questionnaire to all my incoming students. I received a terse reply from one, an established manager of a well-known company. She wrote, "I will not be forced to get up in front of my peers and speak. I have completely avoided pubic speaking and am aware that it has cost me career advances."

I talked with her before class and she told me about an agonizing personal experience. During her senior year in high school, she had spent a harrowing night defending her brother and sister from a raging alcoholic father. When she arrived at school the next morning, she suddenly remembered that she was supposed to give a speech. Not wanting to explain the real reason for her lack of preparation, she asked the teacher for a postponement. The teacher insisted she go to the front of the class. Paralyzed with fear, she stood silently for two minutes before being allowed to sit down. Exhausted and humiliated, she vowed never to make another speech, and she had since avoided public speaking in many ingenious ways.

I encouraged her to trust me and the class and try a brief presentation. "It's okay to be lousy," I told her. "No grades, no judgments; stop when you want to."

She reluctantly agreed. She hyperventilated at first, but then as she recognized the safe environment, she gained more and more confidence. She did a fine job and she was pleased with herself for having taken that big first step. She had broken a link with the past.

One of my executive clients hired me to work with him on a speech. When I started to talk about fears and anxieties, he interrupted impatiently. "Let's skip that," he said. "I never get nervous." I asked him if he had ever had a bad speaking experience. Although he said no, he proceeded to tell how his boarding school headmaster had humiliated him in front of his peers. He swore then that nothing like that would ever happen to him again and he vowed to become an excellent communicator.

Two different reactions to a traumatic situation. One person viewed it as a defeat; the other one considered it a challenge and was driven to practice and excel.

Can you remember an embarrassing or humiliating experience speaking in front of a group? Think back to your childhood. I am amazed when people immediately recall a long-ago situation and describe it in dramatic detail. Their voices reflect the pain. When they've finished, I ask them to go back and create a happy ending. I tell them to relax and visualize the incident ending successfully several times. Your imagination doesn't know the difference between visualizing it and having it actually happen. Gradually, the negative feelings will diminish.

PSYCHOLOGICAL REACTIONS TO FEAR

To understand why fear can have such a tremendous effect on us, it may be helpful to understand how fear affects us physically. Imagine that you are asked to give a presentation and you agree to do so. Let's examine what happens to you physiologically before you speak.

The human stress response is derived from the "fight or flight syndrome" that served to keep our ancestors alive when they were confronted with a saber-toothed tiger. They could choose to face up to the tiger and fight it, or flee.

We have a similar reaction when we perceive a social danger to our ego. Making a fool of ourselves in public is today's tiger. We need to defend our image. As we get ready to speak, our body mobilizes its forces to meet this demanding situation. Hormones are released into the bloodstream to prepare for action. Within seconds, our heartbeat, blood pressure, respiration, and perspiration noticeably increase.

For our ancestors, an increased blood supply to the muscles enabled them to run from the tiger, or battle with added strength. For us, the usefulness of increased muscle power is minimal. We have an intellectual encounter before us. We want to think well, remember clearly, and be creative, quick-witted, and, possibly, funny.

Normally, the brain uses one-fourth of the blood supply. In stressful situations, with the blood rushing to our muscles, the brain is shortchanged. We need to relax our muscles to redirect that oxygen-rich blood to the brain.

Excellent relaxation exercises are stretching, yawning, shaking arms and legs, and swallowing. You can include neck rolls, shoulder shrugs, and sighing. Suck on a mint to help you in the relaxing action of swallowing. Take a brisk walk up and down the hall. These movements help release tension.

Check your body to pinpoint sites of tension. Let's say, for example, you feel tension in your hands. If you can't shake them, visualize all those muscles becoming free and weightless. They will relax but also be ready to spring alertly into action. You want to work toward a state of restful energy. If you don't have a private place immediately prior to your speech, do relaxation exercises earlier in the day or while traveling to the site of your presentation.

Deep breathing will also help you to relax. Inhale to the count of six, hold to the count of six, exhale to the count of six, hold empty to the count of six. Do this several times, and concentrate on the numbers as if they were appearing in the middle of your forehead. Comfortably fill your lungs with air, and let all the tension flow from your body as you exhale. Your heartbeat will slow down and you'll feel calmer. You can do this exercise unobtrusively, even in front of a group.

Incidentally, when we eat or drink, blood rushes to the stomach to help the digestion process, and our brain loses out again. Therefore, it is best to eat or drink lightly, if at all, before speaking. Avoid alcohol.

You Are in Control

Years ago when I taught at a community college, I persuaded the administration to purchase a video camera so I could record my students giving presentations. The first week my students were apprehensive about being videotaped. They said the camera made them nervous. The second week they said the camera made them forget what they were going to say. The third week, a young woman stumbled through the beginning of her presentation, then stopped and said that the camera was making her uncomfortable. "Then please stop," I said. "We will all wait until the camera does something to make you comfortable." Twenty-three students stared at the camera. Finally one observed, "It isn't going to do anything." "I guess *I* made *myself* upset," the speaker said. The class learned the lesson that they had to accept responsibility for permitting their fears and anxieties to affect their behavior.

It is important to remember that a camera or an audience doesn't have control over you. You control yourself and your own tension. You decide if a camera or an audience will stress you out. I know from experience that if you wait for the audience to put you at ease, you will wait forever.

It's a matter of perception. You can either think of the audience as a ferocious tiger and magnify the threat or regard it as a group of eager listeners and minimize the threat.

Someday I'm Gonna Own This Town!

An excellent way to decrease the fear of speaking is to increase your desire to speak. General Chuck Yeager, the famous test pilot and the first man to break the sound barrier, says that he feeds on fear as if it were a high-energy candy bar. He is able to do so because he has learned to welcome challenges as opportunities to test his skill, a chance to prove that he is the best. Think of it as the "Someday I'm gonna own this town!" theory.

Here's how this theory applies to public speaking: Karen Northrup, of the U.S. Army Corps of Engineers, had been forewarned that the general would be attending her presentation. She spent weeks designing slides and also preparing a duplicate backup of viewgraphs. After hours of practice, she felt completely confident that her presentation would go well.

But when the day arrived, the program dragged as speakers went beyond their allotted times. The schedule was already twenty minutes into her segment. Northrup's turn came and she was walking to the front of the room when the general stood up and announced, "I have to catch a plane in ten minutes. Can you conclude your presentation by 4:30?"

Northrup began to panic. "This is unfair," she thought, doubting that she could delete three-quarters of her presentation and still deliver a strong message. What if they were testing her, the only woman in the room? Wouldn't it be simpler to suggest postponing it until another time? But a small voice kicked in and quietly whispered, "You can pull it off and you will look terrific! Go for it!"

Everyone was staring at her, waiting for an answer. She decided to condense the introduction, flash quickly through the first ten slides, include two main points, give examples, and conclude with statistics.

"Certainly," she declared in a businesslike tone. Her eyes swept the room and made contact with the general. "Let's begin by looking at some new information," she said.

Northrup successfully responded to the last-minute challenge by adjusting her presentation to fit the situation. Her fear was put aside as she concentrated on actions that would bring about the response she wanted. She even felt that she won extra points because her audience recognized she handled a difficult situation creatively.

There are few times in public speaking when everything is automatically stacked in the presenter's favor. The difference between the amateur and pro is that the amateur gives up in the face of fear, while the pro considers the situation a challenge and forges ahead. The person who stands up to speak and the person who is too fearful to get up have one thing in common: They both feel fear. But good speakers *acknowledge the fear, harness it,* and decide to *use what abilities they have to accomplish their objective.* Their desire to succeed is stronger than their fear of failure. They espouse the same philosophy as Humphrey Bogart's character in *Desperate Hours.* He mused, "I kinda like taking chances. You don't take chances, you might as well be dead."

Tips on Alleviating Fears

- *Prepare 150 percent.* Preparation is essential and will make you feel more secure. Get started immediately, even if you have a long lead time. Time and effort spent in design will translate into being self-assured during the actual delivery.

- *Acknowledge and accept your fears.* Accept nervousness for what it is: part of the preparation for speaking. Remember two things: First, our bodies initially react involuntarily and in proportion to the perceived degree of danger. Second, we can minimize further bodily reactions by our attitude and self-talk. We can make these reactions work for us.

- *Label your physical reactions in a positive way.* You don't have to label quickening pulse, dry mouth, and sweaty palms as negative reactions. When you're in front of the audience, interpret your feelings as anticipation, exuberance, and enthusiasm. Say to yourself, "I'm very excited about being able to share my ideas."

- *Avoid visions of doom and gloom.* If you have a vivid imagination, you probably have wild fantasies of what could go wrong in a presentation. Remember that if you perceive the situation as extremely threatening, your body will increase the nervous energy it needs to fight this big battle. Be honest. What is the worst thing that could possibly happen? Has it ever happened? Will anyone throw tomatoes at you? Will the audience get up and leave? Will you really fall in a dead faint in front of everyone? By being realistic about the "danger" of the situation, you will have more control, and you will reduce your stress response.

- *Give yourself permission to make mistakes.* Nervousness occurs in direct proportion to how judgmental we are about ourselves. Give up perfectionism; it makes unrealistic demands. If you aren't making a few mistakes during every speech, you aren't taking risks and growing. Calculate how you can minimize your risks. Don't seek to be a keynote speaker at your company's annual convention if you really don't have the expertise or experience. But do challenge yourself. Do a few things that terrify you. Anything less is boring.

- *Don't worry if memory lapses throw you.* What if you forget what you're going to say next? I've had it happen to me. There are several things you can say: "Let me summarize what we've covered so far," for example, or "Would anyone like to comment on my last point?" Audiences like interactivity. Refer to a handout, or go on to the next slide, even if it is out of context. These actions gain you some time to remember. The main thing is not to get rattled. If you maintain your composure, the audience may not even be aware that you lost your place.

- *Trust in your abilities.* Believe in yourself, your ideas, and the value they have for other people. Act as if you are in control, even if you don't feel that way. Remember that you appear much more confident to your audience than you feel. Create mental pictures of your success as if it has already happened. Your confidence will build as you progress.

- *Laugh.* Has anyone said to you, "I laughed so hard that I fell off my chair?" That is because muscles relax when we laugh. I saw a young man hooked up to medical equipment that measured his heart rate. Suddenly he broke out into a hearty laugh and his heart rate shot up to 140 beats a minute. However, it fell dramatically because the laugh released tension. Stretch and do some isometric exercises for relaxed alertness.

- *Make a decision to do your best and then let go of your concerns.* Unwarranted fears actually reduce your abilities by distracting you from information that might be relevant. Relax and spend your time more productively visualizing and anticipating a successful outcome. Laurence Olivier remarked to actor John Mills:

> There's a trick I've used on occasion and I find it works; try it. Go to the theater early on the first night and get made up well in advance of the curtain. Then walk onto the stage and imagine that the curtain is already up and that you are facing the audience. Look out at them and shout, "You are about to see the greatest performance of your entire theater-going lives. And I will be giving it. You lucky people!" Tell them that once or twice. Then go back to your dressing room and relax, and you'll find when the curtain does go up, you'll have the necessary confidence.

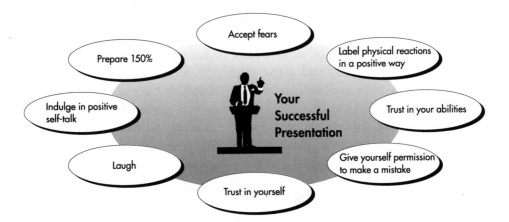

- *Rate the difficulty of your presentation on a scale of 1 to 10.* Your presentation may have important consequences, but think back to other difficult situations you handled well. Acknowledge the critical or doubting voices in your head, but replace them with positive thoughts of times when you were confident and in control.

- *Become totally involved in the moment.* Fear is generated by *conflict:* The body wants to do one thing, the mind says to do something else. The need to succeed and do everything perfectly becomes more emotionally charged when the stakes get higher. Your boss and key decision makers may be attending your talk, and this presentation could propel your career forward (or slam doors in your face). Obviously, you have to be at your best even when part of your mind wants to exit quickly. Make the conscious decision to stay and be terrific. One remedy for fear is to tell yourself, "I would rather be here than anyplace else." It works!

- *Make the audience your partner.* Fear can come from feeling separate from others. It is counterproductive to imagine your audience is weird, has cabbage faces, or is naked. Remember that the audience wants you to succeed. They want you to be good; audiences don't come to hear a bad presentation. If you're comfortable, they will be comfortable. Direct your energy into becoming involved with your audience. Work toward generating thoughts and pictures in their minds. Make an emotional connection.

- *Take every opportunity to speak.* Experience and practice will reduce your anxiety considerably.

Fear Can Be Helpful

President Franklin D. Roosevelt was writing a speech he planned to give on Jefferson Day. The last sentence read, "The only limit to our realization of tomorrow will be our doubts of today." FDR picked up his pen and added, "Let us move forward with strong and active faith." The next day, April 12, 1945, while busy signing bills and correspondence, he suffered a massive heart attack.

Fear can be very helpful. Remember the way you reacted to stress and anxiety by cramming for that statistics exam and acing the final? Take charge. Change your perspective. Use relaxation techniques. Totally concentrate on the moment. Prepare Plans B and C. Suddenly, this whole anxiety-ridden situation is transformed into a challenge.

A doctor doesn't fall apart when she sees an accident victim. A good tennis player doesn't crumble when he faces a strong opponent. A prepared speaker doesn't crumble before a large discerning audience. A small amount of fear can actually push us to perform beyond our accepted capabilities. The challenge is to recognize the fear, accept it, and make it work for you.

Key Ideas

- Understand and accept your fears. Minimize the perceived threat to your ego.

- Identify each fear and develop a strategy to manage each fear.

- Redirect the energy from stage fright into stage presence.

- Do deep breathing and relaxation exercises.

- Make the audience your partner.

CREATING A POWERFUL MESSAGE IN TEN STEPS

CREATING A POWERFUL MESSAGE IN TEN STEPS

"The basic function of using a conscious format for increased creativity within the problem-solving activity of design is to free us from uncertainty, confusion, and other insecurities as we travel along our journey."
—-Don Koberg, Jim Bagnall
The Universal Traveler

OVERVIEW

This ten-step model provides you with a systematic approach to preparation. If you follow this design, you can be certain that you have left nothing to chance. Each of the steps is important. Following these ten steps in sequence will make efficient use of your time and energy.

I asked one of my clients how he divided his time in preparing for his presentations. "I spend about 30 percent of my time in research, 40 percent in actually writing it and creating visuals, and about 5 percent in rehearsal." "But that only adds up to 75 percent," I said. "What about the other 25 percent?" "Oh," he replied, "I spend 25 percent of my time worrying about it before I ever get started."

One of the biggest problems facing any presenter is procrastination. This ten-step system will help you get started immediately. Like any efficient system, this one will require modification with feedback, or as new information is obtained. For example, you may need additional graphics for clarification if you find your audience lacks the expected expertise in your field. Your short-term objective may change because of last-minute information from a colleague. When you give a more interactive presentation, as in a consulting situation, you may wish to delete information on the spot.

If you master this system, you will be able to get the response you want on a consistent basis.

Ten Steps to Effective Presentations

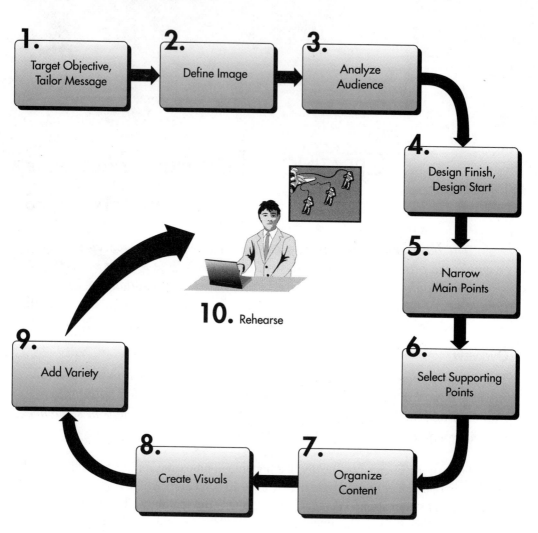

Chapter 4

TARGETING YOUR OBJECTIVE, TAILORING YOUR MESSAGE

"Those who cannot tell what they desire or expect still sigh and struggle with indefinite thoughts and vast wishes."
—Ralph Waldo Emerson

OVERVIEW

If you're going on a trip, you need to know your exact destination in order to plan which route to travel and what to take with you. When preparing a presentation, target your specific objective in order to decide which approach is best, what information to include, and what information to omit. In this chapter, you will learn four criteria for targeting your objective.

Once you have decided on the response you want from a particular audience, tailor a message that will bring about that response. Your message should include the main idea plus the profit-value to the audience for accepting and acting upon your information. Time spent on deciding on your desired outcome will give your ideas direction and focus. It will simplify your preparation and save time, money, and frustration.

During the Middle Ages, knights were trained with innovative "teaching machines." The knight on horseback charged a wooden figure mounted on a pivot. If the knight struck the shield exactly in the center, the wooden figure would fall over. But if it was struck off center, the figure would swing around and hit the knight with a club. That is instant feedback!

It's not necessary to employ such drastic measures for those presenters who stray from their purpose; but many audiences resent a speaker who wastes their time. Target your objective and customize your message to get the response you want. Fail to do so, and you leave the audience response to chance. If you know what you're aiming for, you will be able to evaluate whether you succeeded.

ESTABLISHING YOUR GENERAL OBJECTIVE

Ask yourself: Why am I speaking on this subject to this audience at this particular time in this specific situation?

Is it because I have something worthwhile to say due to my background, credentials, or personal experiences that gives me unique insight? Or is this speech a job requirement in my position?

Your general purpose may be to inform, to persuade, to consult and recommend, to inspire, to entertain, or a combination of these.

You may want to:

- Inform customers, the general public, or investors about your industry, products and services, or research.

- Report progress on a project to management or other disciplines.

- Explain new concepts, goals, or regulations to employees.

- Describe and interpret your research for a professional association meeting.

- Instruct others on:

  — The use of new equipment or software.

  — Methods or policies.

  — Job requirements.

- Persuade others to:

  — Accept and support your ideas.

  — Give you resources: time, personnel, equipment, financing.

  — Accept changes in operational procedures.

  — Start or continue projects.

  — Buy a product or service.

  — Select your company for research or design.

  — Establish an ongoing partnership.

  — Accept the validity of your research.

In Chapter 3, I suggested you consider your audience your *partner*, not an adversary. You're actually asking them to duplicate your thinking and feeling. Take them by the hand and guide them along, step by step, through the maze of information to your chosen destination.

Criteria to Target Your Specific Objective

Once you have decided on your general purpose, decide what *specific* response you want to elicit from your audience. What do you want them to *do* when you finish? Do you want them to take immediate action? Future action? What do you want the audience to *feel*? What attitude changes do you want?

- Gear everything to the listener's point of view. Say to yourself, "After hearing my presentation, the audience will be able to..." or "The purpose of my speech is to have my listener...." Keep a specific purpose in mind and the chances are much better that you will hit your target.

- Use active, concrete words such as identify, contrast, or cooperate to describe the response you want. For example, "When I finish speaking and answering questions, the client will be able to identify or describe three reasons why it is of value for them to purchase my product, service, or idea and they will welcome my suggestions." This objective, worded from the audience's viewpoint, is distinctly different from saying, "I want to sell them a new phone system." These verbs give you specific responses to aim for. Avoid such vague objectives as, "I want my audience to *understand* or *know* about my product or situation." Choose two or three active verbs from the following columns that describe the response you want.

To Inform	*To Persuade*	*To Entertain*
Analyze	Accept	Amuse
Compare	Buy	Enjoy
Contrast	Contribute	Laugh
Define	Convince	Like
Demonstrate	Cooperate	Please
Describe	Disagree	Smile
Explain	Follow	Welcome
Identify	Help	
List	Join	
Plan	Offer	
Repeat	Participate	
Summarize	Volunteer	

- Determine if your results are measurable. If you gave the audience a test, could they actually do what you ask? For example, can they clearly describe to someone else three reasons why they should sign a contract or continue funding your project? Did they sign the contract or give approval to your project? Are your results measurable?

- Have realistic expectations. A common mistake is to take on too broad a subject, one that can't be covered adequately in the allotted time. You will be more effective if you limit your message and cover your points well. It is unrealistic with high-ticket, complex products or services to expect commitment on the first or second sales call. Each visit should complete a positive, progressive step, building a relationship, and moving your prospect closer to the final sale.

One of my clients said that it had taken him three years to get an appointment with a prospective customer, who was satisfied with an out-of-town competitor. My client would have ten minutes to tell his prospect why he should stop doing business with the other company and hire him. What could he possibly accomplish in ten minutes? I suggested changing his long-term objective of getting a contract to a short-term goal of establishing rapport and gaining the prospect's trust. My client decided the prospect needed to recognize two reasons why it would be valuable to consider him and his firm. His message emphasized what clearly differentiated him from the competition: (1) His prospect's expanding company could benefit from his experience and innovative ideas to help them grow their business; (2) his customized, local service would enable him to work closely with their staff. By limiting his objective to selling himself and his uniqueness, he gained interest and a future appointment where he was able to discuss his services.

Focus and simplify your message. It isn't necessary to tell everything about your organization, product, or service. Size your message and the amount of detail to the needs and knowledge of your audience, as well as the available time.

What Is Your Payoff?

Your emotional state is an important outcome. What do you want to feel when the presentation is over? Did you reach your objective if your department agrees to work overtime for a special project, but only because you made people feel guilty? Is there a way to accomplish your objective and maintain a good relationship for future endeavors?

Your emotional objective in a presentation might be to feel that you have been helpful and have efficiently presented your information. You might like to be perceived as caring and concerned. In order to feel good about the experience, you may need to have the audience indicate their time was well-spent. You will feel successful if there are numerous requests for information, an enthusiastic question and answer period, or requests for you to speak again. Target your emotional objective and pay attention to the factors that will help you achieve that objective.

TAILORING YOUR MESSAGE

Your message should capture the essence of your presentation. Be brief. Be specific. One of the best messages I have ever seen was a sign along the dusty road through a wild animal park. It said:

Trespassers Will Be Eaten

Note the distinction between your *objective* and your *message*. Your *objective* describes *the response* you want. Your *message* is *what you say* to elicit that response. Your *message* contains the *main idea plus the profit-value* to the audience.

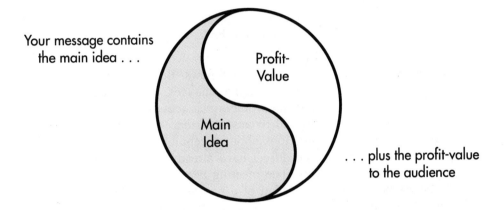

Your message contains the main idea . . .

Profit-Value

Main Idea

. . . plus the profit-value to the audience

The personals in our local newspaper are filled with vague advertisements: "Looking for someone who wants to add excitement to his life," "who wants some laughs," "to share fun times," "to enjoy all the Northwest has to offer," or "seeking a lifetime relationship." But one recent ad stated, "Need date for series of seven champagne and lobster picnics on one of the most exclusive, private islands in the world. Lady should be attractive, bright, and possibly a little daring." This gentleman stated his objective and profit-value succinctly!

Why should your audience invest precious time listening to you? Does the content of your message help them be more productive, make more money, or save time? Tailor your message and profit-value to your specific audience to help you reach your objective.

Ask yourself the following questions:

- What are the goals of my audience?

- What information do they need to reach those goals?

- What obstacles stand in their way of obtaining those goals?

- How can my ideas, products, or services reduce or eliminate those obstacles?

- What innovative ideas, products, or services will my audience need in the future that they may not be aware of now?

The profit-value must be immediate and significant today to break through the preoccupation barriers of your listeners. It's a "FedEx" world. Even lottery officials realize they must offer instant winners. The public is impatient to find out what they've won, just as your audience will instantly assess whether your information is valuable to them. Practice targeting your objective and being succinct when you "headline" your e-mail. If it contains a value to the recipient, it stands a better chance of being read.

Project updates are fairly common occurrences, but they can be stressful when a great deal is at stake. Greg Baron, former project manager at Lockheed, told me his objective for periodic updates was to convince management that all problems on the project were being identified and addressed. He wanted to report that work was on schedule and within budget. His message was that the review board could be comfortable that the project was being handled well:

> Of course, I always like to report good news. But if an unex-pected problem surfaces, my objective changes. Then I need to convince the review board the problem was identified at the earliest possible moment, and request assistance from them. I make several recommendations about acceptable solutions and give the board options, because I have already solicited feedback from key decision makers. The meeting becomes an informed decision-making session instead of a problem-solving session. One of my guidelines for the update is reflected in a sign in my office that says:

> Your Lack of Planning
> Does Not Automatically
> Become My Crisis

Paths to Your Objective

There may be many routes to your destination. Your long-term objective may be to get a contract signed with a large company. But you may not be able to get an appointment with a key decision maker. When you do speak to a representative, your short-term objective may be to have that person identify the value of setting up an appointment with a key decision maker. You want the repre-sentative to clearly relay critical information that will influence a senior officer to meet with you to discuss your product or service.

Begin by writing down what you want your audience to *do* or *feel* once your presentation is concluded. What is the line of reasoning you expect your audi-ence to follow to reach this conclusion? Your objective may change as you do your analysis. For example, a recent roadway widening project in Washington state was offered for bid. The RFP (Request for Proposal) simply called for qualified consultants with experience in structural and civil transportation engineering and environmental permitting. The proposal seemed to emphasize leadership qualifications, past experience on similar projects, and an approach to meet an accelerated schedule. Respondents were to interview the PM (Project Manager). In talking to the client, other needs not on the RFP surfaced. They included an ability to form partnerships with local agencies and stakeholders, assist the city with funding, and meet an extremely tight and compressed schedule. The PM also noted that experience with the city would be very important in the selection process because this project was very visible to the public and the city needed to feel comfortable with the consultants selected.

Be wary of thinking you can disregard your audience's objective and substitute your own. A software executive was asked to demonstrate his product in such a manner that the audience would be encouraged to investigate all the different graphic programs on the market. The demonstration was an excellent way to spotlight the best features of his product for prospective buyers and portray his company as an educational resource. Instead, he continually referred to his product as superior to anything on the market and displayed his logo at every opportunity. The audience was turned off by his commercialism. He not only lost a wonderful public relations opportunity, but also blemished his company's reputation by pursuing his personal objective, even though he knew that it conflicted with the audience's objective. The audience would likely have been sold had he chosen to inform rather than try to influence them to buy his product.

Objectives May Change Within a Presentation

Audiences are not static. They may completely change their attitude as the speech progresses. That is why you must be attuned to what is happening within your audience to accurately target their responses. Have the flexibility to change if you aren't on the right road, or press your advantage if you are. Some people will plunge ahead even if they are going in the wrong direction and their followers have abandoned them.

These people remind me of an American who was visiting Stonehenge in England with a group tour. The tour leader told an elaborate story about the people who built the austere monument. He thought he had an attentive audience. He dramatically waved his arm and finished by saying, "And now, they are all gone."

"When did they leave?" asked the tourist.

"1215," the tour leader replied.

The American glanced at his watch and said, "Just my luck. Missed 'em by half an hour."

Actively interpret the feedback from your audience so you can underscore a point, heighten an advantage, or stimulate them when they begin to stray. If someone is looking at his watch—or worse, shaking it to see if it is still working—it is time to take a detour without losing sight of your objective. One trainer gives instruction for several hours and then stops for a midcourse correction. She asks the participants to write down on colored cards one question that hasn't been answered and one item they want covered during the rest of the session. She reads the cards and posts them on a bulletin board. The participants are asked to remove their card when the requested material has been covered in the class.

If you have a volatile subject, a short-term objective may be to avoid open conflict. In another situation, your feedback may indicate the need to change your objective to keeping negotiations open and scheduling another meeting. Sometimes attaining your objective may have to be delayed while you reassure the audience that they are getting reliable information from a credible source.

Hidden Agendas of the Presenter

We all have hidden agendas. We want our audience to like us, to be comfortable with us, to admire us, to invite us back again. Your hidden agenda at a scientific conference may be to give an excellent presentation that will enhance your personal image and represent your organization well. However, if this preoccupation with *your* desires supersedes the audience's *needs*, it will negatively affect your body language, voice, and words.

One client insisted his objective for an upcoming presentation was to be the "best speaker at his convention." I acknowledged that this was a worthy personal goal, but asked what he expected the audience to do or feel as a result of his message. He reiterated, "I want them to think I am the best speaker at the convention and ask me back again." I tried to convince him that he would accomplish his personal agenda if he focused on fulfilling audience expectations. But he resisted this objective. Nothing ever came of his speech.

A technical consultant from Hewlett-Packard said that he used to give service as a troubleshooter but now must act as a salesperson as well. He wants his customers to perceive his company's support as valuable and his counsel as worth the monthly service fee. He needs to make his customers comfortable with the system they have bought; but his secondary agenda is to recommend additional products as he identifies a need. He generates trust by concentrating on responsive customer service which, in turn, produces sales.

Right Objective, Wrong Message

A local company decided to honor its staff during National Secretaries Week. A committee came up with the idea of presenting the secretaries with chef-style aprons imprinted with the words:

You're a key ingredient to our success!

The response? Most staffers chucked their aprons in the nearest wastebasket, incensed at the symbol of female servitude. One accepted the gift and said it would look great on her husband. Several of them chafed at being linked with Secretaries Week and insisted they were administrative assistants. The objective was well-intentioned, but the message was open to negative interpretations. Seek feedback from your colleagues before your presentation to ferret out any possible misinterpretations of your message.

Same Objective, Different Message

One of my clients was on a trade show tour to four different cities. She had a thirty-five minute presentation and her *objective* was to have dealers purchase CADD (computer-aided design and drafting) software to sell to their customers. Her *message* was that this software would be a profitable addition to the dealer's inventory because so many of their customers—from homemakers to architects—would be potential buyers. She wanted to emphasize that the dealers didn't have to learn CADD. The home office would give the vendors plenty of support.

This sales rep also sold directly to customers. Her objective to sell software remained the same, but she changed her message because of the needs of a different audience. She emphasized that CADD, which replaces the pencil and eraser, would save them time, money, and frustration. They could also get customer service through a toll-free number. Your objective may be to sell your service or product to different clients, but you need to adjust your message for each audience.

Clear Objective, Strong Message

Raymond Hull, co-author of *The Peter Principle* and a frequent adviser, would make me write my objective with a magic marker on a large sheet of paper above my typewriter or computer. He would glance at my objective as he read my writing. If the material rambled from my objective, he tossed it in the wastebasket. I winced when a favorite example or choice quote met its fate, but he insisted that every line had to serve my purpose.

Tentative, lukewarm messages arouse the same kind of tentative, lukewarm emotions in the audience. Bold statements voiced with conviction will engage an audience from the start and keep them alert and attentive.

Does your data or evidence deepen the audience's understanding? Do your supporting points contribute and progress toward your objective? If they don't, eliminate them.

As you go along, and as new facts come to light, modify and make refinements to your purpose. But there does need to be a point when you cease major revisions. An architect told me that once a client had decided on the design of a building, the client had to bite the bullet and give the go-ahead for construction. There could be minor modifications, but if the design kept changing, the building would end up outrageously expensive and would never be finished.

Be precise and clear in your mind about your destination. When the Spanish conquistador Hernando Cortez began his conquest of Mexico, his men balked at storming the unfamiliar beaches and climbing the perilous mountains. Cortez ordered his ships to be burned on the Gulf of Mexico shore in full view of his army. He destroyed them so his troops would concentrate exclusively on their objective of victory.

Once you have targeted your objective and tailored your message, you will know what type of data and evidence you should look for. You will have guidelines to analyze the material you find. The time spent on sharpening your objective and customizing your message will simplify the rest of your preparation.

KEY IDEAS

- Determine the outcome you want for your audience and for yourself.

- Be specific and have a measurable outcome.

- Narrow your subject to realistically fit the allotted time.

- Emphasize the profit-value to your audience in your message.

- Realize that your objective may stay the same, but your message may change.

Chapter 5

DEFINING YOUR IMAGE

"The more you are like yourself, the less you are like
anyone else and the more unique you are."
—Walt Disney

OVERVIEW

People buy you first; then they buy your services, products, or ideas. An
appropriate image will increase your chances of obtaining the response you
want. This chapter will help you make specific choices that build a well-defined,
distinctive image for a particular speaking situation. We will discuss character-
istics of the ideal communicator in scientific and technical fields. You will learn
how to communicate your strengths. You will also learn how your self-image
affects the way you communicate and that the best image is to be yourself.

"Can you make me charismatic?" asked an engineer. "I have to speak to this
committee next week and they really intimidate me." Then he laughed, "Actually, I'd settle for coming across more confident and decisive. Can you do that?"

"I'm sorry," I said, "I can't make you more confident, or more decisive, or
more creative, or give you any of those qualities."

He sounded very disappointed. "I thought that's what speech consultants
were paid to do."

"If I show you how to appear confident and you don't already possess
confidence, I would be teaching you how to pretend. I can show you how to
communicate in a stronger, clearer way those qualities that you already do
possess. You probably have lots of excellent qualities that other people aren't
even aware of."

"Can you think of a time when you felt self-assured, on top of the world, and you were proud of something you accomplished?" I asked.

"Well," he reflected, "I remember a baseball game in high school. I was confident of my ability as a pitcher because I had a no-hitter going as we started the ninth inning. There was a scout present from the Cardinals."

"Were you nervous?" I asked.

"You better believe it! The other team had their two best hitters coming up and everyone in the bleachers was screaming."

"Did the scout intimidate you?" I asked.

"Not really. I wanted to show him he hadn't wasted his time coming to see me. Of course, there was pressure, but I felt I could pull it off. I pitched a perfect game and we won. The scout told me the Cardinals were interested in signing me to a contract. That meant I couldn't go to college. I decided not to sign, but it was a great feeling to be asked."

"We've got something to work with," I said. "I'll help you visualize that situation again and transfer the exhilaration, challenge, and confidence to your upcoming speech. We'll show that committee they aren't wasting their time."

Maybe you haven't pitched a perfect ballgame, but perhaps you can remember a situation in your life when you felt in control, when you had high self-esteem and confidence. It might have been when you completed a difficult project before the deadline, beat your time running a mile, or were complimented by your boss for solving a tricky problem. Visualize that moment again. Immerse yourself in the positive feelings. Transfer those emotions of high self-esteem to your present situation, and your body language and voice will communicate a strong image that helps you get the response you want.

IMAGE IS COMMUNICATION

When we talk about image, we are talking about communication. The moment a person meets you, he takes in thousands of impulses to form a first impression. He selects, interprets, and filters information based on his expectations, his past experiences, the circumstances surrounding the situation, and his self-esteem at that moment. He makes an instantaneous judgment of whether he will trust you or deal with you before he hears a word you say.

Your image begins with your physical presence. You are judged first by the tone of your voice and then by the words you say. Words carry information. But your feelings, attitudes, physical state, and self-image are revealed by your voice and body language.

Studies indicate that more that half the information we receive is from body language, another third is from tone of voice, and less than a tenth comes from the words themselves. Your body language, facial expressions, and tone of voice make up your image. Those three elements must be in harmony with each other for you to communicate effectively.

For example, a woman runs into the room, jumps up and down, and declares in a loud, excited voice, "I won the lottery!" There is no confusion here. Her body language, voice, and words clearly communicate the same thing.

Or you are listening to a paper being delivered at an association meeting. The presenter's words indicate the research is of critical value, but his voice lacks conviction and drones on in a monotone. His body language lacks energy, and he rarely makes eye contact or acknowledges the audience. You have the feeling he would rather be somewhere else. If the presenter is bored by it all, how important could his findings be? When a presenter sends conflicting signals, an audience will select body language as the most accurate message.

Your image isn't necessarily what you are or what you want to be, but how you are perceived. And perception is reality. You may be sending unintended messages. If you are the least bit foggy, unfocused, or inconsistent, you leave room for other people to interpret you any way they want.

Clearly define how you want to come across to others, and your body language and tone of voice will be more in harmony with your words. There will be less chance that you and your message will be misunderstood.

Sell Yourself!

Historically, factual, dry presentations that lack personality have been acceptable in the scientific and engineering communities. But the rules have changed; so has the game. Science and technology are front-page news. There is an expectation now that scientific and technical information will not only be of value to an audience, but will be presented in a compelling fashion.

In today's climate, scientists and technologists must sell themselves. Clients want to talk to the very people who do the work for them, not upper management. The general public demands to know and understand the data that control their lives.

Since many of the topics you discuss are not captivating to everyone, it behooves you to use every means available to engage your audience. They may not be looking for perfection, but there shouldn't be anything to detract from an image of professionalism.

The Ideal Image

In my research over the years, I have asked hundreds of people to name characteristics of the ideal professional image in scientific and technical fields. These diverse characteristics were mentioned frequently:

- Technically accurate, knowledgeable

- Thoroughly prepared

- Organized, structured

- Enthusiastic about subject

- Witty, humorous

- Confident

- Comfortable with subject and situation

- Good appearance—appropriate, unobtrusive clothing.

But three characteristics were mentioned by almost everyone:

- Experienced

- Trustworthy

- Credible.

Experienced

It is extremely important in the scientific and technical field that you come across as being experienced. You must appear to have been "in the trenches." Your company also needs to be seen as having leadership and an excellent track record. Each time you speak, you are in a position to enhance or detract from your company's image. The audience is not listening to you as an individual. Your audience is listening to Mr. Hewlett-Packard or Ms. IBM.

If you are speaking with a potential customer, it is not enough to refer to your performance on previous jobs and how key issues were handled and problems solved. Arnold Silver, known for his work in superconductivity for TRW, said, "There is a current need in technical fields to be perceived as *more than competent*. A customer must believe you have the ability and intelligence to solve his problem."

It is comparable to being introduced to a potential date at a cocktail party. Will it benefit you to recite what wonderful things you did for a previous companion? Or is the person you meet more interested in how you will behave with him or her? In the theatrical world, the reviews of your last play aren't as important as whether you can sell tickets for the current one.

Perhaps you were proficient using a software program last year, but the upgrade leaves you in the dust. The rapid updating of technology mandates you are perceived as someone who can *adapt his experience to the present situation, keep up with frequent changes, and deliver exceptional-quality products and services.*

The people buying your product or service may not fully understand it, but they must be absolutely convinced that you are proposing the right solution for them. Simply put, people like to do business with people who know what they are talking about. Become an expert in your discipline and keep up-to-date with your audience/client's industry. Cultivate your image as an educational resource, and you will find others rely on you for advice.

Do other people perceive you as experienced? In addition to your words, people look for self-assurance, strong presence, and direct eye contact. Your tone of voice is important. There should be no hesitations. Your voice must convey authority. You need to appear comfortable with yourself and with your material.

Trustworthy

A presenter must be perceived as having integrity. Jim Collins, business consultant and former faculty member of the Stanford Business School, defines integrity as "Telling the truth, fulfilling your commitments, living up to your word, and not cheating others—even if you can get away with it." Collins makes the distinction between being ethical and having integrity. "Having ethics means having values or a moral code. However, having integrity is withstanding the pressures of compromising that moral code."

Today, when the actions of so many business executives, clergy, and public officials have been revealed as corrupt, it is important for your audience to be able to rely on you and your information. If agreements are made between you and your audience, your audience needs to believe you will keep your part of the bargain. Investor Warren Buffet advised his son, "It takes twenty years to build a reputation and five minutes to ruin it. If you think about that, you will do things differently." You will find the cost of restoring your good name to be exorbitant.

A trustworthy person has open body language and appears genuine. He has direct eye contact. He projects a feeling that he will follow through with whatever he says he will do. There's no evasiveness or hesitation in his speech. He is relaxed and sincere. Phil Theibert, a speech writer for a large western utility, advises:

> There are only two elements that make a speech great; sincerity and brevity. Tell the audience what you really believe in your heart. If you're not sincere, if your soul doesn't come through in that speech, you will hear snoring from the front row all the way back to the exit door. Remember: Keep it simple, keep it plain, tell them the truth, and get the hell out of there.

Credible

Expertise in your field and *trustworthiness* give you an image of *credibility*. A CEO of a software company told me, "Investors do not put their money into products or services, but they do invest in people. That is why professional image is so important. People buy you first." If I perceive you to be an expert and I perceive you are fair and reliable, then I will be more apt to believe you.

The presence of more women in technical and scientific professions has helped to create acceptance, but women still need to project an authoritative image to avoid being stereotyped or misinterpreted. One woman in sonar technology sales wears a lab coat when she demonstrates equipment and talks about her twelve years as a technician to emphasize her credentials and to gain trust. Rules change and women are unsure whether they should be feminine, masculine, or androgynous. Many women feel pressure to be something they are not.

I asked a man why he felt that a certain woman had risen to the top of her field. He replied, "Her expertise and her presence. She looks comfortable with herself. In our industry, such an attitude is an indication to me of professionalism." What he is suggesting is that women can be themselves without relinquishing their authority and leadership qualities.

An awkward and nervous delivery can undermine the expertise you wish your audience to acknowledge. A strong presence will enhance your credibility and influence others to accept your message.

CHOOSING APPROPRIATE CHARACTERISTICS

Make a clear choice of the image you want to project for your next presentation. Select five adjectives that describe this image. You know that you want to appear to be experienced, trustworthy, and credible. Other qualities that you might want to communicate are:

Knowledgeable	Empathetic	Compelling
Enthusiastic	Patient	Attentive
Sincere	Composed	Innovative
Confident	Organized	Consistent
Caring	Prepared	Reliable
Committed	Realistic	Resourceful
Fair	In control	Supportive
Open	Adaptable	Decisive
Futuristic	Unflappable	Independent

A young engineer in one of my classes at Westinghouse remarked, "How can I come across as knowledgeable when I don't know everything about my product? Besides, I look much younger than my age of twenty-six. Youth might be an asset in the computer industry, but it's a strike against me when I'm selling high ticket items on complex projects."

I asked him to come to the front of the room. I told the class that if you are young or look young, people will question your experience and intelligence. You will need to compensate by having impeccable content and superior presentation skills. Naturally, you should have as much product knowledge as possible. However, being knowledgeable simply means *knowing where to get information.*

I had a class member ask this young engineer a difficult question. I instructed him to appear bewildered and answer, "I don't know." He shrugged his shoulders and looked down as he answered. He looked uncomfortable.

"When unexpected questions come up," I told him, "remember that your listener gets most of his information from the body language and tone of voice. Keep your posture aligned, make eye contact, and answer 'I don't know' in a strong voice. Tell them you will get the information to them as quickly as possible. In this way, you can still convey that you are a knowledgeable person even though you don't have a specific answer to the question." He tried it again, and the class agreed he came across as intelligent and concerned. Another suggestion is to plan ahead and ask a more senior person if you can defer questions to him or her. If you do, have it understood that you would like a brief answer and then you would like to assume control again.

I asked another engineer to name one of her strongest characteristics. She felt that she was organized; feedback indicated that her audiences didn't think so. When we videotaped her, I noted she rambled when she spoke and it was difficult to follow her thoughts. She also became involved with her visuals and excluded the audience. We worked on simpler visuals and I encouraged her to establish frequent eye contact with her audience. I advised her to introduce one section with, "Let's divide this into two areas of discussion," or refer to her ideas as "Number one, number two." In addition, I suggested she give a short preview of her presentation and also make a bulleted list on a visual. This way, both she and her audience had an outline to follow. This situation demonstrates how one's perception of oneself may not coincide with the audience's view, and the important need for feedback.

Be aware that your strengths are also your potential weaknesses. For example, many engineers and scientists feel a need to be precise. Although this is a desirable trait, it can be carried to unfortunate extremes if a presenter gives detailed, minute statistics that overwhelm and confuse the audience. Competency in a subject is a strength, but a presenter can be so knowledgeable that she inhibits creative feedback and involvement of the audience. Moderation is the key.

Be Real

> *"I always wanted to be somebody, but I should have been more specific."*
> —Lily Tomlin

When Ralph Waldo Emerson first heard the abolitionist Wendell Phillips speak, he wrote in his journal, "The first discovery I made of Phillips was that while I admired his eloquence, I had not the faintest wish to meet the man. He had only a platform existence and no personality." In other words, Emerson did not perceive Phillips as genuine.

The best image is a real one. When I talk about projecting an image, I am not suggesting that you fake one. Your image should not be a deception or a veneer to conceal who you are, but rather it should accentuate your best qualities and strengths. Think of image as a tool for communicating your competence and credibility because it will underscore everything you say.

The more you are aware of your own identity, the more powerful you will be. We shouldn't need to rehearse to be ourselves. But many times, we think we aren't salable the way we are and we assume a facade.

It was thought-provoking to hear the story about a reporter who gushed to one of Hollywood's most famous stars, "What's it like to be Cary Grant? Everyone wants to be Cary Grant." And Mr. Grant replied, "Hell, I wish I *were* Cary Grant. I became the person I wanted to be and he became me or we met someplace in the middle." Various stories reported that Mr. Grant was an insecure person in search of an identity with which he could be comfortable.

Roger Ailes, communications consultant to many top U.S. executives, stresses that:

> Once you reach a comfortable, successful level of communication, you never have to change it, no matter what the situation or circumstances or the size of the audience. The key element is that you not change or adapt your essential "self." *You are the message* and once you can "play yourself" successfully, you'll never have to worry again.[1]

Have courage in your convictions and trust in your common sense. Be careful about following every best-selling author who claims to know the perfect formula for instant success. If an author tells people to put a pumpkin on their heads and waltz around their office or lab, there is bound to be someone who will try it.

I always ask participants in my technical presentations classes to give persuasive speeches about their favorite causes. We videotape these presentations. Suddenly, body language is transformed as they try to persuade the audience to adopt their beliefs. They have energy, their voices have more variety, and eye contact is strong. The video reveals that they can be compelling. Then we work on transferring that enthusiasm and energy to their on-the-job presentations.

I was called in to work with a financial officer in a technical company who had to deliver a speech to an accountants association. His expertise was immediately evident, but his delivery was bland. He objected to exercises that I hoped would loosen him up and allow more of his personality to show through. "I'm not interested in learning all those creative techniques. I just want to get up and have something worthwhile to say."

I asked him to name someone he considered to be an excellent speaker. He mentioned an executive. When we started to analyze why the speaker was compelling, he had to admit that the executive was enthusiastic about his subject, was comfortable in front of a group, involved the audience, and used humor and other creative techniques to convey his complex material.

A Sense of Theater

When I suggested to one of my technical clients that a presentation is a performance, he protested by saying, "I'm not an actor, nor do I want to be! I don't want to change my personality. I want to come across naturally, as myself."

I told him I had no intention of changing him. Rather than packaging him, I preferred to unwrap his best qualities. But he needed to take an inventory of his assets and liabilities and determine which assets could be communicated in a clearer way.

The best image is a genuine one, enriched with a sense of theater. Without a sense of the dramatic, you may have *efficient* speaking but not *effective* speaking. It's the difference between getting the audience to *understand* what you say and getting them to *do* what you want. Dynamic, purposeful, energetic speakers can propel their beliefs and ideas with efficiency and ease and break through the preoccupation barrier. Stimulate your audience to get involved. It is important to connect energetically and emotionally.

Self-Talk

The most important elements affecting your communication are your self-image and your self-talk. No matter what techniques you use, if there are negative thoughts running around in your head, they will destroy any attempts to project a strong positive image. If you don't like yourself, your negative body language will overshadow any positive words.

A presenter who is uneasy with himself appears weak. It is important to know and be comfortable with yourself. My mother told me about Jack Kennedy's presidential campaign train coming through our small town in upstate New York. When he spoke from the back of a railway car, she was captivated by his energy. The mesmerizing quality that he had comes through in videos years later. Kennedy had total delight in himself and in the job of the presidency. When someone walks away from your presentation, do they believe you like yourself and your work?

"Humble Doesn't Play Well"

—Jackie Gleason

An IBM engineer specializing in artificial intelligence asked me to help him with an acceptance speech. He had been chosen engineer of the year by an international engineering fraternity.

I stopped him after he'd read the first few words of his speech.

"What is the image you wish to project to the audience and the board who selected you?"

"I think I should be humble," he said.

"If you spent hours on a committee to select the most distinguished, young engineer from the entire country, would you be happy with a humble person who confessed, 'I don't deserve this.' What did you feel when you were notified of the award?" I asked.

"Me? Wow! That's terrific! I'm honored!"

"That's what you should project," I counseled.

I explained to my client that humility is necessary for a speaker, but it also needs to be balanced with confidence. Would you like it if a person opened a gift from you and then said you should find a more worthwhile recipient? Wouldn't you prefer to see a person who enjoys the recognition of such a prestigious group, and who exhibits strong leadership qualities and confidence in his work?

The engineer realized he didn't have to present a false image of himself to curry favor with his audience, but rather could express his true feelings. When he accepted the award, he told the audience about the many years of disappointment and failures, and the nights spent in research and experiments. Then he said, "On any one of those discouraging nights I never would have imagined I'd be standing here accepting this award. This makes it all worthwhile. Thank you for recognizing me and my work."

Sir Laurence Olivier, acclaimed in his lifetime as the world's greatest actor, astutely acknowledged that, "The most difficult equation to solve is the union of the two things that are absolutely necessary to an actor. One is *confidence,* absolute confidence, and the other an equal amount of *humility toward the work.* That's a very hard equation."

Sir Olivier's words rang true when I watched Barbara Walters interviewing the charismatic Michael Flatley, who choreographed and stars in *Lord of the Dance.* Walters stated, "You've called yourself the world's best dancer." "I am the world's best dancer," Flatley declared. "That shows a pretty big ego," Walters countered. Flatley took Walters' hand and walked her to the front of the stage. "18,000 people will be attending tonight's performance. Do you think they want me to come out and apologize for my dancing?"

Feeling Comfortable

Have you ever lost a contract to a competitor and felt you never were told the real reason? You and your team spent weeks doing research, employed state-of-the-art visuals, and had an outstanding reputation in the field. The written proposal was excellent and professionally bound. You knew you were on track because you were a finalist. What happened? What was the deciding factor?

An engineer told me that his department head received a phone call from a prospective client. "We were impressed with your ideas and appreciated all the time and effort you put into research for the proposal. However, we've decided to work with another company," the client informed the department head.

"Could you give me some feedback on why we weren't chosen?"

"We felt more comfortable with their team," was the only answer.

"We were shocked we didn't get the contract," the engineer told me. "We have more years of experience. I know our bid was lower. Our oral presentation seemed to go well. More comfortable—what does that mean?!"

Effective speakers, no matter how technically expert, must project human qualities that show others they are *easy to deal with*. This translates into being comfortable with someone. It includes listening skills, responding well under pressure, a sense of humor, and body language, dress, grooming, and conduct appropriate for your profession.

"We're interviewing several speech coaches," I was told by a prospective client, "but it all comes down to a personality fit." She added bluntly, "We'll either like you when you come through the door, or not." It is said that 25 percent of your audience will love you, 25 percent will hate you, and 50 percent will have a "show me" attitude. You can't do anything about the 25 percent that have no rationale for their dislike (you may remind them of a cousin they loathed), so concentrate on the 50 percent that you can persuade.

WHAT TO AVOID

The image you project of yourself and of your work, department, or company should reinforce positive traits and avoid perceptions that could have an adverse affect. What image do you want to avoid?

A project leader should not appear out of control, disorganized, or incompetent. A salesperson should not be perceived as insensitive, but as caring and concerned about customer problems. Even though your knowledge may be superior to that of everyone in the audience, an arrogant or abrasive attitude will turn off your listeners.

More people are fired because they can't get along with others than because they lack technical expertise. Individuals don't make it up the career ladder if their personality conflicts with their peers or superiors. I have seen brilliant people in line for top jobs get passed by because they had a "short fuse" and couldn't control their anger.

One employee was called the "Grim Reaper" because every time he came in the office, he was gloomy. It was difficult to deflect his negative energy. Are you positive, upbeat, and have a touch of humor in your communications? You will be welcomed with open arms if others perceive you as someone who will not be unduly upset by daily hassles.

We live in a stressful, unpredictable world, and we aren't always feeling in top form, rested and focused. Consciously choosing characteristics you wish to project will be especially helpful in those impromptu moments when you are asked to say a few words in front of an important committee or unexpectedly meet a potential client. Your body language and tone of voice will reflect your predetermined image and will help you achieve your purpose.

KEY IDEAS

- Recall a time when you felt high self-esteem and transfer those emotions into the present situation.

- Seek to know yourself and be genuine in your communications.

- Project an image of being experienced and trustworthy to be perceived as credible.

- Be aware that your strengths are also your potential weaknesses.

- Be easy to get along with.

Notes

1. Roger Ailes, *You Are the Message* (Homewood, Ill.: Dow Jones-Irwin, 1988).

Chapter 6

ANALYZING YOUR AUDIENCE

"If you know the enemy and know yourself,
you need not fear the result of a hundred battles. If you know yourself
but not the enemy, for every victory gained, you will also suffer a defeat.
If you know neither the enemy nor yourself, you will succumb in every battle."
—Sun Tzu
The Art of War

OVERVIEW

You want your audience to listen attentively, therefore it is important to find out what is of value to them and what they want to avoid. This chapter explains a system that will enable you to determine the level of knowledge of your audience. It will help you zero in on the information most valuable to the audience. You will be able to demonstrate that you understand their needs, desires, and goals. You will be able to anticipate how the audience will react to your material and to you. You will be able to select appropriate material for your presentation and organize it in the most effective manner.

It has been accepted in the technical, scientific world that facts speak for themselves. But sometimes those facts don't say anything to nonscientific or nontechnical audiences. Scientific and technological terms are often viewed as a foreign language. The public is unfamiliar with the vocabulary. It might seem too difficult to translate the complex technical concepts into something useful in everyday life. And facts can mean different things to different people. In that sense, each of us speaks a separate, individual language. Presenters must speak the "language" of their audience if they wish to avoid misunderstandings, misinterpretations, or boredom.

Effective communication that seems effortless is not accidental. In interviews with model communicators in scientific and technological fields, I found that they take responsibility for the audience's understanding of their topic. They realize it is not what they say but what each listener *hears* that is important. They take time to find out how they can be on target.

Model speakers know their audiences are inundated with overwhelming amounts of information and that it is critical to present only what is relevant and of value to their listeners. You can create strong, clear, concise messages. You can command rapt attention if you take the time to analyze your audience, respond to their needs, and speak in a language they can easily understand.

GETTING STARTED

If you are like most people, you delay working on a presentation until the last moment. I devised the audience analysis checklist from twenty-seven years of coaching clients. The first few questions are factual and easy. You can start working on a plan for your speech the minute you are requested to speak. And if you can answer all the questions, the speech is practically written. The checklist gives you the security that you've covered all the bases and you can concentrate on involvement with your audience.

Do you have a presentation coming up in the next few days or weeks? Read the audience analysis checklist to see how many questions you can answer immediately about your prospective audience. The first few questions are obvious but essential. The rest of this chapter will expand on the checklist and explain in detail how the answers should affect the way you write your speech. When you finish reading the chapter, review the checklist and start the detective work to find answers to as many of the thirty questions as possible. Every time you have a presentation, use the checklist to guide your audience analysis.

DETECTIVE WORK

Before you begin gathering information or writing your speech, make contact with someone responsible for the meeting or event. Find out who is going to be present and obtain background information about the audience. Your liaison usually is the person who originally requested you to speak, but it could be someone in the same department or a program chairperson. Perhaps you are making a sales call or presenting a proposal to a group or committee. Request phone numbers or e-mail addresses of audience participants and inquire about their high-priority concerns. Ask the group for a few of its annual reports and newsletters. Make a trip to the library for industry magazines. Find out if there is a Web-site address connected with the organization and download information about their latest news items, new services and products, investor reports, or staff changes. Conduct an on-line search for industry related items.

AUDIENCE ANALYSIS

Situation
1. Requested topic _____
2. Name of person, group, client, department, etc. _____
3. Liaison's name _____
   Phone #: Work_____ Home _____
4. Address or location of speech _____
   Room #_____
5. Occasion _____ Date _____ Time _____
   a. Business meeting _____ Formal _____ Informal _____
   b. Principal speaker_____
6. Meal_____ Refreshments _____
7. Attendance: Voluntary_____ Required _____
8. Other speakers _____
   Topics _____
9. Length of presentation _____ Q&A _____
10. Title of speech _____
11. Person introducing you _____

Audience
12. Size of audience _____ Men %_____ Women %_____
13. Age levels _____
14. Occupations _____
15. Educational levels _____
16. KNOWLEDGE of the subject _____
17. UNKNOWNS to define or explain _____
18. PROFIT-VALUE or GOALS of AUDIENCE _____
19. FIXED BELIEFS and ATTITUDES because of professional, social, religious, departmental, cultural affiliations _____
20. EXPECTATIONS of the group _____
21. Specific CONCERNS of audience _____
22. Who is the DECISION MAKER or KEY PERSON(S)? _____
23. Any recent event, situation, local color that you should take into consideration? _____

24. What example, story, personal anecdote, historical reference, humor will "bond" you to the audience? _____

Speaker
25. General attitude toward you:
    Known _____ Unknown _____ Friendly _____
    Hostile _____ Indifferent _____ Show Me _____
26. Perceived credibility on subject: High _____ Medium _____ Low_____

Content
27. Precisely what do you expect your audience to do or remember?
    When I finish speaking, I want my audience to _____
28. My message (must include profit-value) is _____
29. What three main points must I make to inform or persuade the audience to my point of view?
    1. _____
    2. _____
    3. _____
30. What emotions should I elicit from my audience to get the response I want? _____
    _____
    _____

I often send out a short questionnaire to be completed and faxed or e-mailed back to me by several people who will be in the audience. That questionnaire can have variations of the questions on the checklist. For example:

Concerns: "What specific question do you want answered in the meeting?"

Level of Knowledge: "Have you ever used a similar software program?"

Attitudes: "Would you install equipment for solar energy in your home?"

If you are speaking at a meeting, ask to be sent a copy of the agenda. Request an advance program for a formal conference. Sometimes final programs are not known or printed very far in advance. Ask for e-mail or phone numbers of other panel members. Don't apologize for asking a series of detailed questions—a group should feel flattered that you intend to complement the rest of the program.

FAMILIARITY BREEDS CONFIDENCE

This is the way George Novak, of NASA Lewis Research Center, approached his monthly in-house progress reports:

> The objective of the meeting is to report the status of the project to the center director and his staff. If any activity on the critical path slips or is behind, it can cause problems in the whole system. The meeting is held in the administration building and I am familiar with the room. There are usually ten to fifteen people present, depending on who's in town. It's mandatory for the director of the project and his deputy, but voluntary for the rest of the staff.
>
> Since it's a small group I don't use a microphone and I know that I will be using viewgraphs for visual aids. I can speak fairly fast but if it was a larger group, I would have to slow down my delivery.
>
> There are two other presenters. The program manager gets up and gives a technical analysis and the Safety and Quality Assurance manager gives a report. Prior to the meeting, we discuss at length what we each will be saying to avoid any surprises.
>
> Normally it is from 1:30 to 4:00 P.M., right after lunch. I always eat a light lunch, and I get physically relaxed by walking the three-quarters of a mile to the administration building.

Prior attention to details will put you at ease and give you more control. If your presentation is at another site, make sure that you get the liaison's name and phone number both at work and at home. You may need to check with him or her for any last-minute changes in the audience, situation, or location, or if you have a personal emergency.

YOUR AUDIENCE'S MAP OF THE WORLD

A painter, geologist, and rancher were all looking down into the Grand Canyon. The painter said, "I can't wait to come at dawn and paint the sunlight breaking over the horizon and the incredible mix of colors." The geologist remarked, "Look at the different strata. I want to go down and collect some ore samples." The rancher gazed down the vast chasm and exclaimed, "That's a heck of a place to lose a calf!"

Everyone has his own "map of the world" or unique world view. To understand people and to connect with them during a presentation, we must recognize that they do not respond and behave according to our perceived "reality" of a situation, but to their perception of "reality." Two scientists working together in the same field who have had different educational and work backgrounds and have read different books may be worlds apart in communicating because their perceptions, vocabulary, and images are different. Difficulties are compounded when communicating across organizational lines such as from research and development to sales and marketing.

Recently I attended the open house of an architect. He is a patron of the arts, and beautiful paintings and sculptures were on display throughout his new offices. In the reception area was a print of the famous painting, *The Doctor*, by the Scottish artist, Sir Luke Fildes. The masterpiece shows a country doctor at the bedside of a very sick child. The crisis is near. The doctor looks puzzled and gravely concerned. The mother has her face buried in her arms sobbing in desperation. I paused to look at the exquisite colors and precise detail. A medical doctor was standing next to me studying it. "What do you think, Doctor?" I asked. He peered at the painting closely, stepped back, and announced, "Acute appendicitis!"

Any audience you address will make associations and call up images based on their unique backgrounds, experiences, and knowledge. Could I read your speech and be able to write a profile of your audience? Model communicators make an extra effort to thoroughly plan and prepare. They know that if the audience fails to understand, the reason is not that the audience is stupid, but that they failed to present the information in ways the audience could readily comprehend. One presenter said, "I'm there for understanding, not to convince them how smart I am."

MIXED AUDIENCES

Gender Differences

The number of women continues to grow in science and technology. Be aware that the men and women in your audience will hear and interpret your information in a distinctive way, just as they each would use different language to describe the same situation. If you want the full extent of your message to reach both genders in your audience, you will find it worth your time to educate yourself about gender differences.

For example, some male presenters still predominantly use sports analogies. Even though many women are avid sports fans, presenters should select metaphors that have meaning for everyone in the audience. Women are also more likely than men to actively signal their attention by nods and "mmhmms." Men tend to interpret such actions as signifying agreement with what they are saying; however, the woman's intent may be to show that she is listening and comprehends, not necessarily that she agrees.

Gender differences can also be found in nonverbal communication. Conventions of posture, facial expression, tone of voice, and use of pauses differ between the sexes.

Avoid using sexist language. Rephrase statements to include men and women or use neutral descriptive terms. For example, say "workforce" instead of "manpower," "supervisor" instead of "foreman," and "artificial" instead of "man-made." For help in using nondiscriminatory language, consult various available references.

Generation Differences

Many social scientists describe the over-fifty-five generation as traditionalists. They are used to information being delivered at a slower pace with someone guiding them along the way. The "baby boomers" in the forty-five- to fifty-five-year range usually need options. They created the salad bar and twenty kinds of toothpaste in five different flavors. The twenty- to thirty-year-olds are said to be "challengers," and don't exhibit as much loyalty to tradition or to their employers. They will probably question everything you say. They are used to searching out information by using faster, multi-threaded, hyper-linked delivery systems.

However, you will find a multitude of exceptions to the above classifications. I know one sixty-year-old CEO who refuses to have a computer on his desk, and another, older CEO who insisted on personally designing his company's Web page. Most audiences have a wide age range, and it is best to remember that individuals in these audiences will respond to different logical and emotional appeals. It is important to remember that emotional links to the experience of one generation cannot be taken for granted in another. You may be treading a very fine line in your choice of examples. Use analogies, references, and facts to call up images familiar to the entire audience, but try to include some examples that will touch the time frame of different age groups.

Most high-tech industries are populated by young employees in key positions. These young buyers have specialized backgrounds and less experience than their counterparts of 15 to 20 years ago. Savvy salespeople realize they must act as an advisor and educate these clients. These younger buyers are more cautious and take longer to make a decision. The Advertising Research Foundation reports 35 percent said they took up to 90 days after the initial inquiry, 28 percent said they waited between three and six months, and 19 percent said they placed an order after six to 12 months. Patience and persistence will be rewarded.

What Do They Really Want or Need to Know?

If you are selling technical equipment, you will speak before individuals with very different needs. Flexibility is important, since the focus of the audience can be technical or business oriented. An audience can consist of engineers, architects, computer wizards, business executives, security analysts, consultants, prospective buyers, or current customers.

For example, the chief executive officer and upper management of a company are concerned about profits. These executives aren't usually occupied with software details or computer technologies but want to know whether your system is a good business investment. They need to understand your product in terms of costs and benefits: Will this system allow them to meet their corporate objectives?

A bio-tech scientist gave an excellent internal presentation to a diverse group of marketing, administrative, and sales personnel. Instead of spending the majority of his time on the history of the project or his detailed experimentation, he gave a brief overview of his methodology and then immediately focused on his interpretation of the results, recommendations for next steps, and a proposed timeline for follow-up experiments.

He recognized that his audience had no interest in details of failed experiments. They were more interested in possible products and resulting profit for the company.

One of my clients was speaking on digital imagery to a national scientific conference being held at a local university. I called to find out who had registered for the session. Over 40 percent of the attendees were college students. I cautioned him not to dismiss this portion of the audience as insignificant, as some might do. These were his future customers, employees, and investors. He needed to anticipate and respond to their needs and goals. Many of these students would have little knowledge of the topic, and would also want to learn about career prospects in this industry. My client adapted his more sophisticated presentation by providing some basic information, giving a CD demonstration for illustration, and including comments about the future outlook for products and companies in this field.

At technical conferences, the research papers presented are intended to increase the audience's knowledge even if the information may not be of immediate use. Audience members will watch the clock or may doze off if the speakers are too obtuse, or fail to relate how the knowledge will be of value.

Tech-Speak

Today the scientific and technical world is being enriched by an infusion of different cultural perspectives. Your audiences will increasingly be made up of the multilingual people for whom English is a second language. At Digital Equipment Corporation's Boston plant, for example, 350 employees speak 19 different languages. Identify with your listeners and consider the difficulty of understanding complex, technical information in another language. Choose your words carefully and use universal associations that will evoke accurate images in the minds of your listeners.

It is essential to explain unfamiliar terminology. Joe Warner, a district manager for Compaq Computer Corporation, was addressing a nontechnical audience of writers on the use of various word processing programs. He began by issuing a challenge to the audience and putting himself on the line: "If I introduce a concept, a word, or an acronym without giving you a concrete example or explaining it in your terms, raise your hand and I will give you a quartz clock." Everyone was eager to catch Warner, but he only needed to give out three clocks during his ninety-minute talk. Not only did he ensure the audience's attention with his pledge, he let the audience know that he cared enough to speak in their language. Would you feel comfortable repeating a similar offer?

One technique Warner used was to break down groups of unfamiliar words, such as "file allocation table."

> All of you are familiar with files in a file cabinet. The word allocate means to distribute, and you can allocate your information into different files. As writers, you know that a table of contents helps you find information in a book. Now you know what a file allocation table does. If you walked into a library, you would need a card catalog or some system to find out where certain books are located. Your computer has to know where information is located. A table of contents or file allocation table asks the disk to find it for you.

Specific Concerns

What are some specific views held by members of the audience because of professional, departmental, social, religious, or political affiliations? Find out some group commonalties and concerns, but be mindful that each person in the audience also brings individual attitudes that need to be considered.

A U.S. Navy spokesperson addressed many different audiences concerning the controversial subject of building a Navy homeport for a super aircraft carrier in Washington state:

- Realtors—who wanted to know about the influx of population in regard to housing availability and sales.

- School officials—who were also concerned about the increase in population, but in regard to their services and operational budgets.

- Emergency services—who were mainly interested in police and fire services, sewers, and electrical power for an increased population.

- Political groups—who wanted to know the exact plans and the funding to be provided by Congress.

- Engineers—who were curious about technical details of the proposal, such as the confined aquatic disposal technique.

The Navy spokesperson sent out a questionnaire to determine the knowledge and composition of each group, requesting the answers in writing. His careful audience analysis revealed that each group required different information and a unique approach, even though the basic topic was the same. All of the groups except the engineers required translating the technical terms into language understandable and useful to each audience.

Remember, you do not allay the fears of your audience by giving them information they already know. You can eliminate fears by clearing up misconceptions and half-truths and inaccuracies, but you first need to determine their present state of mind.

Perhaps you own a small start-up company and wish to give a presentation to prospective investors. The investors know the grim statistics of the high risks and the failure rate. Your job is to convince them that your products or services will survive the competition and adapt to the massive changes in the marketplace. Can you convince them that your management has controls in place to minimize their risk and maximize their profits?

Underlying Values and Attitudes

If your audience believes that your profession is only out to make money or blindly pursue the latest technology without concern for the client or the societal impact of transaction, then you need to acknowledge these feelings and start from where they are before they will even listen to you. You can't ignore deep-seated resistance if you wish to persuade or influence them—gradually build credibility with facts, information, and recognition of their concerns.

Rick Daniels, of Waste Management of North America, Inc.'s Oregon subsidiary, was facing a very hostile audience. He was the project manager on a proposed landfill in a small town in Oregon. (It is now the third largest landfill in the United States.)

A community meeting was called. His only prior knowledge about the town was that it was undecided on the landfill. He wasn't prepared for the twenty people that showed up. Rick said he felt that he was in front of a lynch mob ready to hang him. They yelled at him, "No landfill, no trash! Are you going to live here? You're just an outsider sending your smelly trash up here!"

Rick said, "I didn't give a presentation. It was time to do an on-the-spot analysis. So I asked them about their fears and concerns and I listened. I took notes and asked more follow-up questions. At the end of the meeting, I told them that I wanted to make sure I understood what they were disturbed about. I read back my notes—they wanted to know about litter, traffic, smell, and so on. I got agreement that I had accurately understood and recorded their questions. Then I read them the conditions under which they would agree to having a landfill. I told them, 'We are committed to doing this in accordance with your wishes. Stick around—I'll be back in two weeks when I have the answers.'"

Two hundred people out of a town of 454 showed up at Rick's next meeting and he patiently started through his list item by item. It was a free-for-all exchange of ideas, but Rick described it as a very positive meeting. He would be back again to continue to answer their environmental concerns and also to tell them the economic value that the project would have for the small town. By the fifth community meeting, when Rick had addressed every worry and fear and given them adequate guarantees that the local environment would not be adversely affected, not one person spoke in opposition.

The presenter who is aware of the audience's attitude can avoid provoking or increasing hostility and can incorporate the audience's shared values effectively.

Audience Expectations and Beyond

When the audience heard or read that you were speaking on a subject, what were their expectations about your presentation? How did the meeting announcements, publicity, and newsletters describe you and your proposed talk?

I learned a famous scientist was coming to a local university and called to find out his subject matter. I couldn't understand the title and the person on the phone couldn't explain what it meant. I decided to take the time and make the effort to go, but I finally left after listening to thirty minutes of a lecture that had nothing to do with the author's books or field of specialization. I can respect the fact that he may have wished to speak on something other than his world-famous research, but the title and advertisements should have made this evident.

You will have much more control over the expectations of your audience if you prepare a short paragraph that succinctly describes your talk and submit it to your liaison. Everything may not be used for publicity, but at least there is a better chance of the basic details being accurate.

Once you have determined how best to fulfill the expectations of your audience, you should decide how to incorporate the unexpected into your speech. How can you be innovative? You can establish your objectives and then ask the audience for additional areas of concern or interest. Write down and promise to address these concerns during your talk. You need to be in command of your topic, but the audience should immediately feel you understand their needs. At one presentation, I stood at the door and, as I greeted the attendees, I asked them to submit questions about their day-to-day communication challenges on 3x5 cards. It gave me time to think about the answers and also a chance to gauge the responsiveness and energy level of the audience.

Fulfill the expectations of your audience and then have something unexpected. You might use wit and humor in your serious speech. You might have colorful handouts, unusual animated graphics, or a completely different approach to your topic that surprises the audience.

Model communicators know that a dramatic presentation aids in retention of data. Will anyone ever forget Microsoft's introduction of Windows 95 with fanfare, videos, music, celebrities, commercials, and spectacular advertising?

Finding Common Ground

What example, story, personal anecdote, or historical reference will bond you to the audience? When Rick Daniels faced hostile townspeople about the proposed landfill, he told them that although he didn't live in the town, he was a fellow Oregonian. He was a native of the state and he was as concerned about environmental issues as they were.

Sometimes it is difficult to find common ground with your audience. One day I got a phone call. A voice drawled, "You are getting shoved down our throats. We don't want you coming down here to teach us to articulate. Stay home!" Well, I didn't.

I opened my sales seminar by saying, "Last night I heard you talk about hunting. I used to be very comfortable with a rifle. I also went fishing all the time and had a dog, rabbits, owls, and five raccoons as pets." The audience members began to sit up and listen because they were all hunters. "When I was a teenager," I continued, "I had my own trapline for muskrat and mink. I'm here to teach you to make better sales presentations and more profits for this company. We can swap some hunting and fishing stories at the coffee break." And we did. I could relate to them and talk their language, so they decided to give me a chance. They worked hard on their presentations and they turned out to be a fun group!

However, a word of warning...a common bond must be real. If it is forced, you will come across as a phony and do yourself more damage than good.

Getting to Know You

Your credibility will depend on whether the audience perceives you as competent and experienced. What or whom do you represent? What is the general attitude toward you? Are you known or unknown? Are you the best person to give the presentation? Are you viewed as an authority in regard to your topic? Is the subject so technical that you will need an expert with you to answer specific questions? Does the audience perceive you as closely allied to their values? Is the audience friendly, hostile, indifferent, or does it have a "show me" attitude? Unless you are a well-known figure, most members of your audience will expect to be shown that you deserve their attention.

KEY DECISION MAKERS

Is there a decision maker in the group? If you are selling a large computer system to a company, will it be to your advantage to focus on the engineer in the group? Will the engineer make the final decision or will the CEO sign the check? Is there more than one decision maker?

ANALYSIS OF THE KEY DECISION MAKER

1. Who is the key decision maker? _____

2. Is there more than one key person? _____

3. How much do they know about the proposal? _____

4. How will they view the proposal? (financial background, marketing, etc.)
    _____

5. What have they really asked you to present? _____

6. How will they react — some people react differently in a group setting than one on one? ___
    _____

7. What has been the fate of similar proposals? _____

8. Will the proposal reduce the power or influence of the decision maker? _____

9. Will it have any impact on "pet projects"? _____

10. Has the decision maker been on the record as opposing such a proposal? _____

11. What is the decision maker's style? Does he or she want all the alternatives spelled out? Lots of analysis, computer print-outs? Graphs and visuals? Quick-action steps? More committees and studies? _____
    _____

12. Do others have to be persuaded? Get consensus? _____

13. What is your fall-back position — if any? _____

14. What is the next step? _____
    _____

15. What criteria must be met? Budget, time, personnel, equipment? _____
    _____
    _____

What is this person's style? Does the decision maker want all the alternatives spelled out, lots of analysis, computer printouts, graphs and visuals, quick-action steps, committee work, or studies? We all employ various styles of behavior, but there is usually one governing style:

- An *analytical* person will expect facts, trust facts, and remember facts. Be prepared with additional convincing details if he or she suggests delaying a final decision.

- A *socializer* who is more intuitive and creative will be more comfortable with hypothetical concepts. Socializers have large egos—beware of taking away any sense of their power.

- A *compliant, amiable* type may be easier to convince, but get a commitment from them to follow through. Include information about how your proposal will affect the people involved.

- A *dominant, driver* type of personality will be results-oriented and want you to immediately get to the substance of your presentation. Do your homework and don't have any frills!

Robert W. Lucky, Corporate Vice President of Applied Research at Bellcore, cautions that "Key members of the audience can be laugh leaders and emotion leaders, and skeptics, and sympathizers. A few can go a long way to swaying the effect on the whole reaction to your presentation." Gain rapport with these individuals as well as the ones in charge of facts.

Use the checklist to analyze key decision makers. Your answers should influence your approach to your presentation.

- Be aware there is a built-in bias against giving up the status quo, as this may indicate that things were done wrong before.

- Major changes meet more reluctance than minor changes.

- It is sometimes best to address major concerns of the key decision maker up front so that he or she knows you are aware of problems.

ADAPTING TO THE AUDIENCE

What if your lead time is so short that you cannot do a thorough audience analysis? Even if you have one hour, you can still run through the audience analysis checklist to formulate an approach for dealing with an unfamiliar audience. If you get in the habit of using the checklist and considering the variables in each audience and situation, you will find even impromptu speeches easier to write and present.

What if there is no way you can conduct an audience analysis? This is normal for many training situations. John Moore conducts five-day classes on specialized software at Hewlett-Packard with employees he has never met. He constantly seeks feedback and monitors his audience. He tells the class, "My first objective is that you have a good time and have fun this week. My second objective is that you learn something. You don't have to memorize everything I say. It's okay if you don't know or understand the concepts in the beginning— that's why you're here."

He asks numerous questions to determine their level of knowledge. He makes eye contact after presenting an important concept and often he repeats a concept. He only goes forward when new information has been assimilated.

Moore says he "senses" when the class is confused. If they become very quiet or write noisily, he gives more examples to explain a complex command. He is very responsive to body language and the "energy" of the individuals. He is never buried in his material but is totally involved with checking responses and selecting the best way to coach his students.

ACTION STEPS

Here's a suggestion. As soon as you schedule a presentation, select a bright-colored folder with a pocket. Write across the top with a large, black, felt pen the audience, date, subject, and title. Fill in as many questions as you can on the audience analysis checklist provided earlier in this chapter. You might want to copy the checklist from the book and enter it in your computer. Start collecting anecdotes and statistics and put them inside the folder or copy them to a file on your disc. Do your detective work and complete the checklist. You will have a revealing audience profile—their "map of the world."

This profile will give clues that will influence your choice of objective, the organization of your presentation, and the style of delivery. It will aid you in relating your exact message to the needs and beliefs of your audience and will help you get the response you want on a consistent basis.

KEY IDEAS

- Determine the profit-value of your objective to the audience.

- Find out the level of knowledge of your audience and start from where they are.

- Realize that everyone has a different "map of the world."

- Fulfill expectations, but then do the unexpected.

- Anticipate audience reaction to you and your message, but stay flexible.

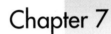

Chapter 7

DESIGNING YOUR FINISH FIRST, YOUR START SECOND

"A speech is like a love affair. Any fool can start one
but it takes considerable skill to end it."
—Lord Mancroft

OVERVIEW

In this chapter we discuss how to make compelling first and last impressions.
The finish of your presentation should be planned first. Design your conclusion
so that your final words will move your audience to acceptance or action. What
they hear last, they remember the most. This chapter also suggests ways to help
you get off to a strong start and involve your audience with your opening words.

FINISH FIRST

Plan your finish first. It may seem strange for you to begin with your ending, but
your objective and finish determine and prescribe the body of your presentation.
Every runner knows that races can be won or lost in the final lap. By the same token,
your final words can be the deciding factor that determines your ability to reach your
objective. A savvy runner plans the race and saves energy for the critical "kick" at the
end. You too should plan and pace yourself in order to make a strong finish.

Your finish is not your last main idea. It is a separate part of your speech that ties
everything together. A well-thought-out game plan can anticipate and minimize the
following hazards.

Avoid:

- Introducing new ideas in your finish. This may confuse your listeners. A
  lawyer never brings up new evidence in a closing argument.

- Rambling or fading away. Don't leave critical, persuasive points up in the air.

- Confusing the audience with unclear action steps. Uncertain audiences won't do anything.

- Misjudging your time. You don't want to rush, condense, or eliminate your finish.

- Going overtime. This is a thoroughly unprofessional tactic and one that won't be appreciated by the audience or other speakers.

Your finish should:

- Restate your message.

- Heighten the emotional connection.

- Reinforce why this information is of value.

- Summarize the use of the information.

Follow these four strategies as you prepare your concluding remarks:

Wrap things up with a quotation that restates and reinforces your main theme.

You can make some final comment about the quote in your own words. Carol Bartz, Chairman and CEO of Autodesk, Inc., finished her speech to the National Coalition of Girls' Schools' and Technology Conference by saying:

> The educational issues girls face today are going to become job and leadership issues for women tomorrow. There's an ancient Chinese saying that goes, "A thousand mile journey begins with the land under your feet." I'd like to add a twist to this. If we don't give girls solid ground to stand on, they may never gain the footing they need to make the journey at all.

A quotation can refer back to opening statements and answer questions you raised in the beginning. One client began his presentation to the Sierra Club by emphasizing the necessity of providing for proper land usage. He ended his presentation with an eloquent passage from a 1971 documentary about the native Americans' reverence for the land:

> Teach your children what we have taught our children—that the earth is our mother. Whatever befalls the earth befalls the sons of the earth. Man did not weave the web of life; he is merely a strand of it. Whatever he does to the web, he does to himself.

By using this quote my client restated his theme of everyone's responsibility to preserve our planet, and the rhythm of the words had a finality to them that sent a clear signal that he was concluding his remarks.

Heighten the emotional connection.

It is important to heighten the emotional connection between you and the audience, as well as to reinforce the logic of your message. Your final words will strongly influence the feelings of the audience toward you and your message, and their desire to take any action. Are you challenging them, increasing their desire, putting them at ease, or stimulating them to increase their productivity? You may want to challenge your listeners to take action so the future will be a promising one. Stimulating the imagination of your audience can help you reach your objective. Showing how their work is going to be easier in the future will grab everyone's attention.

Andrew Grove, chairman of INTEL, summed up his presentation to the Asia Pacific Information Technology Summit by asserting:

> It is really fairly simple. You need home and business PC penetration, you need to have those PCs connected to the Internet, and you need low-cost available local telecommunication services as well as long distance telecommunication services.

> Regions that will have that will very rapidly be able to avail themselves of the system-wide economic benefits of Internet commerce. Regions that don't will be disadvantaged relative to those that do in a growing, growing and growing divergence.

> The key message I would like to leave for you is we're heading toward a world of a billion connected computers. In a world of that mass of broadly available connected computers, problem after problem can be addressed easier with that technology. Information technology is not a luxury in a world like this. It is nothing short of a competitive necessity. And the way out of the difficult economic times for a company is new products; for an economy is new ways of doing business.

Restate why this information is of value to the audience.

If you are speaking at an association meeting, you may want to confirm your audience's feeling about the worth of their profession and the values the association upholds. In sales, it is important to build a relationship of trust with your prospective client and, therefore, reinforce the idea that you will be there to personally service and support your products. A chemical engineer was selling chemical processing equipment in the detergent industry to a foreign client. He finished his final sales presentation by reinforcing the products' benefits to the client. The client, he said, could expect better performance and lower maintenance costs by using this equipment. This would enable the client to manufacture a better quality of detergent for its customers, which would lead to higher profits. And he assured the client that superior technical support would be available.

Summarize the use of the information.

Ask your audience for an order, an action, or a change of belief. It may be a challenge for the group to act. Many times the audience knows what you want them to do, but doesn't know the first step. Make the "What do I do next?" step or plan of action explicit.

One client's request for action wasn't clear at the end of her presentation. In the following question and answer period, the audience members got so caught up in the intriguing interchange that they all left debating the points of the Q&A without taking any action. You can encourage your audience to act by advising them: "I strongly recommend we adopt this plan by voting 'yes.'" "We can promise delivery in two months if you sign the contract today." "I believe that you will agree that we should allocate funds for further research in this area." Or one engineering company executive who finished his oral proposal and simply said, "We want to work with you. We want this contract!"

You may assess a situation and decide that flexibility in your call to action may meet with a more favorable response. For example, I talked to several members of the legislature who said a pet peeve was the obstinacy of various groups who would demand a certain sum of money for their cause. One Congressman said there was only so much money available for countless worthy causes, and this "all or nothing" approach had a higher chance of being rejected. He said a more realistic proposal would be: "Give us $200,000 and we can complete the project. If you give us $125,000 we can accomplish our first two goals." Your audience may be more comfortable if they make their own choices.

Attention to these four goals will help you shape your ending. Succinctly and effectively bring all the loose ends to a crisp close. Even if you are in the midst of research and do not have conclusions to present, you can still summarize what has been accomplished so far, and announce the next steps and your goals for the future.

Strive to make your concluding remarks as brief as possible. Finish two minutes early. Alert your audience to your concluding words and they will be more attentive and responsive. Former President Ronald Reagan once commented to the Governors Association:

> Well, I've gone on long enough. You know, there's a story about Henry Clay, the senator from Kentucky who was known for his biting wit. One time in the senate, a senator in the middle of a seemingly interminable speech turned to Clay and said, "You, sir, speak for the present generation, but I speak for posterity." And Clay interrupted him and said, "Yes, and you seem resolved to speak until the arrival of your audience."

Start Second

"Life is uncertain. Eat dessert first."
—Unknown

Time has become our most precious commodity. The days of long-winded presentations that take forever to get to the point are no longer tolerated. During the opening remarks, audiences now ask, "What's the punchline?"

Emerge from the starting gate with a purposeful, dynamic beginning. Let your audience know immediately what they will gain from your speech. Suspense has its place in mystery books, movies, and theater but is not appropriate for scientific and technical presentations. Author Robert Gigonore says, "I feel I need to grab the reader by the throat with my first paragraph. When I can say you are mine, you'll have to read the next and the next and finish the book." While I don't advocate physically strangling your audience, I strongly recommend making your first few sentences provocative and engaging.

An information specialist told me that his first question is, "Does the presenter know what she's talking about? Is this another time waster? I want to know what's new. I came to hear her insights, her conclusions and how it applies to me. Don't waste my time!"

Your start should avoid:

• Establishing a fake bond.

• Using humor only because you think you should.

• Creating expectations that you can't fulfill.

• Being too humble in front of a special audience.

• Apologizing for anything, except if an audience has been kept waiting or if equipment fails.

• Complaining about your short lead time or lack of time to discuss your subject.

The goals of the start are to:

• Get attention and rapport with your audience.

• Establish your credibility.

• Announce your intentions in terms of subject, purpose, scope, limitations, and plan of development or approach.

• State the profit-value of your subject to the audience.

Use the following strategies as you prepare your opening remarks:

Break through the audience's preoccupation barrier and capture their attention.

Involve the audience as soon as possible. Ask a question, have them write in their handouts, get them to laugh at a story, or make a dramatic statement— anything to interact with the audience within the first *ninety seconds*. An engineer told me that when his firm makes a proposal to prospective clients, they try to convey in words and visuals "we understand your issues and we have the answers" in the first three minutes. Professor Don Jardine began his speech to an American Society of Training and Development convention with the provocative pledge, "I have not come here to bring you drink, I have come to make you thirsty!"

Louis W. Cabot, chairman of the board of Brookings Institute, captured the interest of his audience with opening words that unequivocally set forth his objective:

> My purpose is to erase any complacency you have about science
> in your lives and to replace it with a sense of urgency and alarm.
> I believe man will do more to shake up the human race in your
> generation and the next 10 generations than in all the 100,000
> generations of man that have gone before us. And we are not
> prepared for it....

> Even in periods of great aversion to it, science does march on.
> And like it or not, the affairs of man can only be managed by
> people who have the skills and concepts of a quantitatively
> trained mind and the competence for scientific, critical thinking.
> People who don't know how to work things out, who are not
> quantitatively and scientifically literate are at the absolute mercy
> of people who are.[1]

If you can tie your beginning into the remarks of a previous speaker, do so. Find something to ad-lib about the department, the group, or the occasion before starting your memorized opening statements. This demonstrates you are in tune with this specific audience and the situation. You will show that you care about the audience and are in control.

Even at an in-house meeting, the presenter's first few words should get everyone's attention focused on the subject at hand. Your audience members may be absorbed in a prior work situation or be thinking about a personal matter.

A humorous story can unite the audience and set a friendly mood. A self put-down can bring a smile. Combine humor with a reference to the occasion or setting, or pay a sincere compliment to your audience and their expertise. One client began a speech to a group of hospital managers by saying:

> I asked David how many would be in the audience today. He
> told me about two hundred-odd people. Well, you were right
> about the number, David, but I must say I've had a chance to talk

> with several of the mangers and they're not strange at all! In fact,
> I have been very impressed with their knowledge and their deep
> concern for their employees and the hospital patients.

If your story is relevant, makes an appropriate point, and you feel comfortable using humor, use it. It will make you seem positive, approachable, and in control.

Sell yourself and establish credibility.

Credibility needs to be established with your opening words, body language, and tone of voice. This is particularly true if a presenter is an outsider to the profession of the audience members. Your audience needs to perceive you as trustworthy and an expert in your field, so they can safely place their confidence in you and your material. Carefully select words and ideas that will reflect your experience and indicate your concern and commitment to your topic.

Demonstrating common ground with your audience will build rapport and credibility. What do you share with your audience? President Jack Kennedy established common ground with a group of civil servants in Albany, New York, by his greeting, "Mayor Horning, Congressman O'Brien, and fellow government employees…." Are there geographical associations or common origins that would have significant meaning in regard to your credentials or topic? Did you attend a local college or can you relate a positive experience in the city? If you are an outsider, mentioning your friendship or former working relationship with someone within the organization can help to establish a common bond. Find some way to indicate that you are *familiar with the audience's world.*

One speaker said she had been told that the key decision maker attending her presentation was aloof and impersonal. The speaker was warned to get down to business without delay. However, she found out the woman was an avid golfer and worked the subject of golf into her opening remarks. She felt she established rapport immediately and her information was more accepted.

Preview your material.

Tell your audience what is coming and give them a blueprint to follow. If you tell them that you are building a condominium and outline what steps you will go through, they can take your facts and ideas and build them in the same way you put them together. If they don't have any direction or plan, they can end up with a pile of bricks or a rickety bridge. Curtis Crawford, President of Microelectronics Group of Lucent Technologies, began his speech to the Trendsetters Luncheon at DePaul University:

> I will be looking at some major trends in global communications
> and their broad impacts, focusing on the emerging multimedia
> revolution and the Internet explosion. My perspective will be
> that of a high-tech industry leader in communications equipment
> and systems, Lucent Technologies.[2]

An in-house presentation doesn't need a detailed preview, but you do need to set a level of expectation. You might make one or two statements to bring listeners up to date on what has happened since the last meeting. You may simply choose to begin with, "Today, I wish to focus on two new procedures," or "At our last meeting we outlined the problems of.... Since then I have spoken with several staff members and I would like to present my recommendation," or "My purpose is not to discuss why the staff has been reduced, but to talk about how the work will be redistributed."

Sell your subject by emphasizing the value of your presentation to this particular audience at this particular time.

Tell your listeners why they should pay attention. How will your information make their job easier and more satisfying, save them money, give them power and control, or improve the quality of their personal and professional life? Today, competition is no longer about features, it's about time. The product/service that can save someone time will prosper. Capture the essence of your message, avoid too many details; your audience will be thankful.

An enthusiastic photocopier salesman asked a friend of mine a few pertinent questions about his profession. Then the salesman introduced his product by saying, "Let me show you how this copier will save you time, money and service calls. You can't afford downtime in your business." My friend listened intently.

One project manager said that he tells a reviewing board in the beginning of his presentation how a specific project is going. That way, the audience doesn't have to wait two to four hours to find out the project's status. He said this gives the board some control. They can say, "Skip the graphics detailing the good stuff and get to the areas where there are potential problems." He believes this is the best use of time for both presenter and audience.

Scientists and technologists usually state their results up front, and then proceed to prove their points and show how they reached their conclusions. This is an efficient technique. Audience members are more willing to give careful attention to complex material if they don't have to start out for a destination blindfolded.

FINISH STRONG, START STRONG

Audience attention is highest at the start and the finish of a presentation. Take advantage of this. Your opening should signal that you are enthusiastic about your message and comfortable speaking in front of a group. Therefore, the audience can be at ease, yet stimulated and eager to hear something of value. Wrap up with a strong finish that clearly asks for the response you want and motivates the audience to take that action.

Now that you know where you are going and how you will begin, you will find it easier to start planning the route you will take.

KEY IDEAS

- Write your finish first as a guide to the direction and focus of the body of the speech.

- Announce your objective in the beginning of your speech and preview the pattern you will follow.

- Tell your audience in the beginning what their profit-value is for listening to your message.

- Make an emotional connection with your listeners in the beginning and again in your ending.

- Have a strong finish. Your audience will remember most what they hear last.

Notes

1. Louis W. Cabot, "Sci-Humanists Unite," *Vital Speeches of the Day* (May 15,1989).
2. Curtis Crawford, "Major Trends in Global Communications," *Vital Speeches of the Day* (April 10,1997).

Chapter 8

Narrowing Your Main Points

*"Nothing astonishes men so much as
common sense and plain dealing."*
—Ralph Waldo Emerson

Overview

*Your presentation should give the audience detailed information, but you also
want to give them the benefit of your insight, your analysis, and your recommendations. In this chapter, we discuss the analytical and intuitive ways in
which you can gather information for the body of your presentation. Step back
and get a fresh, new perspective on your subject. Finding sources for information is no longer a problem; however, it is easy to get lost in the research phase
and lose sight of your objective. You will save time if you force yourself to be
selective and to categorize your information as you collect it. Each piece of
material must contribute to and help you progress toward your objective and
that of your audience.*

You have targeted your objective, defined your image, and analyzed your
audience. You've planned your finish and your beginning. You know how you
will grab your audience's attention right from the start. Now you are ready to
gather the material that develops your message and advances your objective.
You can create your main points by breaking your message into subdivisions.

Your objective determines what your main point will be: what to emphasize,
what to leave out, and what to develop in detail. Choosing your main points is
similar to packing a suitcase for your vacation. If you don't know where you are

going, you might include fishing equipment, a tennis racket, swimsuit, books, casual clothing, formal attire, and so on. But if you know that you are headed for San Francisco or Paris to attend a round of formal parties, you will skip the blue jeans and fishing pole and take evening dress.

REDISCOVER YOUR MATERIAL

Make an attempt to rediscover your material. Adopt the point of view of someone in the audience: step back and look at your subject from a different perspective. When I make a video or a film, the camera often starts with a close-up. It might be a tight shot of two people talking. You can't tell much about the environment they're in because you only see faces. Then the camera pulls back for a medium shot revealing a man and a woman sitting next to a campfire. There is a wagon train and you begin to get more of an idea about the scene. The camera pulls back for a long shot; you see a prairie and a wide river, and the cavalry is charging over the far hill. You now have an entirely different perspective and perhaps an entirely different reaction.

Sometimes we never step out of that close-up shot in our communications. Our egocentricity can severely limit our ability to influence our audience. When a scientist or engineer is too close to her information, she may forget to supply the basics. The audience will lack a frame of reference to understand the main points.

One of the most common mistakes is taking on too broad a subject. Zoom in on your topic and present more in-depth information or step back and give an overall view. Michelangelo didn't say to his model, "I'm sorry that I eliminated your head, but I didn't have enough room for it on the canvas." An artist calculates the dimensions of the canvas and either paints a small area in detail or paints a large area from a distance. Speaking is also an art form, and a presenter must consider limitations of time, space, and situation.

Imagine a target. In the center is a bull's-eye. That bull's-eye represents your main points, the *20 percent of your speech that will make 80 percent of the impact.* For you to achieve your purpose, your audience must thoroughly understand these central points.

Surrounding the bull's-eye on the target is a circle that represents the 20 percent of your presentation devoted to the start and the finish. Around that is a larger circle that stands for the 40 percent of your speech that includes your supporting points. The final circle represents the 20 percent made up of the additional details, frosting on the cake, that are helpful but not essential. If you have to cut your presentation short, you could easily eliminate this final ring and not destroy the core of your presentation.

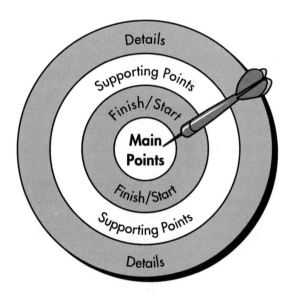

BASIC RESEARCH: ANALYTICAL/INTUITIVE

"Basic research is what I'm doing when I don't know what I am doing."
—Werner von Braun

Beginning research can leave you feeling overwhelmed, but remember: the best way to climb a mountain is step by step. Taking the first few steps will start those creative juices flowing.

There are two approaches you can use to gather material for your presentation. One is intuitive, the other is analytical. The *intuitive* approach uses brainstorming techniques, free association, branching, or mind-mapping ideas instead of a linear outline. We know from research that the brain relates more efficiently to information when it is organized in patterns than to the same information in sequential form.

One way to free-associate ideas is to write your topic, such as "Existence of Life on Other Planets" in a circle at the center of a piece of paper. State your objective under the circle, which might be "to convince my audience that life did exist on Mars." Recall anything you know about your topic and start writing down words and thoughts; draw pictures and symbols for at least ten to twelve minutes in any pattern that naturally evolves. You might cluster pictures and words relating to organic molecules; group other ideas about mineralogical features observed in the ancient meteorite found in the Antarctic, conditions necessary for these small single cells to live such as temperatures, proof that water existed, etc. Your finished picture might reveal the scientific experiments on the Delta Flight. This exercise will undoubtedly stimulate ideas you haven't considered and will suggest innovative approaches to your information. Bouncing your basic concepts back and forth with another person can be helpful.

Here's another exercise to engage your intuition. Turn off the screen on your word processor so you won't be able to judge your ideas. Type as rapidly as possible for ten to twelve minutes on your subject matter. Even if you have nothing to say the first few minutes, don't quit! When you turn on the screen, you will probably be amazed at some of the perceptive ideas that have materialized.

You might even try some of the computer-based "thinking software" to help you generate ideas. These programs ask you specific "what if" questions, reverse your objective, encourage metaphors, and gently but persistently goad you into new avenues of thought.

If you are part of a team, participants can be linked electronically with laptops and brainstorming software. An overhead projector shows attendees' input on a screen—all under the safe cover of anonymity. Ideas and comments can be ranked, outlined, categorized, and archived for later perusal.

A good time to think about your topic is before you go to sleep or when you wake up. Some people free-associate better when they are jogging, walking, driving a car, or taking a shower. Keep a notebook handy to jot down notes. Talk into a tape recorder. Don't try to judge or analyze your thoughts, because this can limit your creativity. You are looking for a variety of ideas.

The *analytical* approach is a systematic search for relevant facts and data. The journalist's questions of Who? What? Where? When? Why? and How? are good starting points. For instance, ask yourself these questions about a process or an object:

- *What* is it?

- *Who* uses it?

- *Where* is it used?

- *When* is it used?

- *Why* is it used?

- *How* does it work and *how* can it work most efficiently?

If your topic is about an object, additional questions might be:

- *What* is it made of?

- *How* do the parts relate to the whole?

- *What* is the synergistic effect of components in the system?

Start your research with basic facts you already have available. The Internet has forever changed the way we research data. Many encyclopedias, indexes, and databases are available to you from your office or home computer; however, many are available only in libraries. You will, therefore, probably find it useful to seek out the services of a good librarian. On-line research services give you access to the world and will help you locate text references you need from millions of scientific journals, technical reports and patents, as well as chemical thermodynamics, materials science, and biomedical research documents. You can interact with universities, teachers, other professionals in your field, association colleagues, and take part in Web chats, forums, and e-mail exchanges.

One of my clients was preparing a presentation on the latest developments in biomaterials. Her objective was to have her audience understand the basics— from raw materials to products, markets, and potential risks associated with this technology. Gathering material for her presentation required using a broad range of current information resources—personal files, books, journal articles, legal information including patents, Internet resources, and a community of experts.

1. Starting with her own database and files, she reviewed relevant notes, comments, and articles she had gathered on the topic over the last few years.

2. She also browsed the Internet discovering Web sites which provided related information. Some of this information was obtained through news services and Web sites for biomaterial companies. Some of the information she decided needed to be verified by another source. She also found a few listservs which she decided to monitor.

3. In the library, she used the on-line catalog and looked up printed material. She found two books written on biomaterials which included substantial bibliographies. The bibliographies provided some leads for other information. Since most information in books is at least a few years old by the time it is published, she also used journal articles for the most current information.

4. Her first task was to master the language about this subject so she could use these keywords to conduct searches on different databases. Every field and subfield has its own language. The encyclopedia and the two books already in hand were very useful in identifying several concepts which could be used as searching terms. She started out searching several databases accessible through her personal account with an on-line information service. She found an article written by an international expert and several other abstracts that looked promising. She had the option of downloading articles and storing them for later perusal and printing, or having them printed out immediately. The database source would also mail a copy of the material to her through the on-line document order system. But she wanted to keep the expense down and went back to the library and looked up the printed references provided by the databases.

5. The librarian also helped her get a book through interlibrary loan from an academic library located in the region.

6. In the library she also logged onto other databases which provided business information (Business Index), patent information (CASSI), and medical information (Medline). All of these sources provided her with appropriate citations since her subject was interdisciplinary.

7. At an association meeting, she heard of a paper by the keynote speaker and asked if she could have it sent to her.

8. The final part of her research process was conducting personal interviews over the phone and by e-mail with experts identified in her literature search. In some instances, her personal network was just as valuable as the databases. She found out that it was necessary to prepare for each interview and know when and where she was going to use the information. Otherwise, it became a stimulating chat, but the information was not useful. She took notes, but taped interviews were more accurate. Most people did not object to being interviewed if she asked beforehand.

As you gather your material, evaluate and classify. Then make a list of active verbs you could use for accomplishing your specific objective. "When I finish talking," you might say, "I want my audience to be able to *identify* my main points, *restate* or *explain* them, and *compare* or *contrast* them to other information." By using these verbs you can qualify the material as you collect it. If you want to *convince* your director to send you to a conference, you'll want to find information on this year's topics and presenters that prove they are relevant to your latest project. Gathering information on previous years' agendas or the many tourist attractions of the site city will not help win a travel voucher.

Many speakers set up individual files on their word processor and then store quotes, ideas, and other pertinent information. If your topic is "Interplanetary Voyages," use one file to store information on Mars, another for propulsion and communications systems, another for the long-term effects of space travel, and so on. You can import and merge information between files as you begin to organize the information into clusters and themes and put it in sequential order. Each file would list all the sources and contain information about that aspect of the subject.

One scientist told me that he couldn't work exclusively on a word processor. "The advantage of serendipity is lost if you just use a computer," he said. "You can only retrieve whatever you remember you stored away. If I don't remember what I put in, I'm not going to find it. In order to know what I have said, I print out all the information so that I'm not limited only to a screenful of information."

Set Limits on Your Research Time and Process

At some point, you have to bite the bullet and begin a final outline or actually write out your presentation. If you discover gaps, you can always go on the

Internet and do more research or conduct further interviews. Authors George and Meridith Freidman in *The Intelligence Edge,* remark, "Information is cheap if you don't care how long it takes to get it. But time is expensive and each successive unit of time purchased tends to cost more."

Get rid of dead weight. Will this particular bit of information help reach your goal or should it be eliminated? If the point, no matter how fascinating, does not logically relate to your objective and move the presentation along, eliminate it. Continue this weeding-out process right up to your final rehearsal.

The scientists who spoke at the NASA Lewis Research Center Business and Industry Summit needed to help businesses *identify opportunities* for long-term partnerships with Lewis. The presenters had volumes of information about their research, but were scheduled to speak for only twenty minutes. The difficult task was to condense their research into a few minutes so they could spend most of their time emphasizing benefits and profits to their audience.

Perhaps some of the material that you feel needs to be included, but isn't quite essential, can be documented and given as a reference. A handout can supply your audience with resources, statistics, abstracts, or a bibliography. Supply a Web page address that will give future updates or an e-mail address to answer questions.

DEVELOP YOUR MAIN IDEAS

Help Your Audience Remember

Can you recall all of the main points from a presentation you heard last month? Or an important conversation last week? What was the essence of a phone call yesterday?

Twenty-four hours after your presentation, your audience will have forgotten 75 percent of your material. For that reason, you should repeat and reinforce the main ideas (the 20 percent in the bull's-eye) in different ways to help your audience retain the information you have prepared.

When you have all your information assembled, list the three to five main points the audience needs to know at the end of your presentation. Your audience will remember the points better if you speak in triads or groups of three such as three points, three examples, or three sets of numbers.

John McDonald, president of British Petroleum Oil Company, in a speech to the City Club of Cleveland, realized the value for narrowing his topic:

> To keep the topic today to manageable proportions, I intend to
> focus my comments only on one important element of the
> debate, the impact of automobile emissions on air quality. There
> are three main areas of current concern arising from convention-
> ally fueled automobiles and they are air toxics, carbon monox-
> ide, and smog.[1]

Each of your main points should carry the same weight and relate to the whole picture. If a fourth point in the above example was about motorists who neglected to have frequent oil changes, it would not have had the same importance as the other three points.

Just as you might begin assembling a picture puzzle by grouping the colors, start to group the ideas you find under different headings for your main points. How do they interlock? What are the dominating relationships? You may quickly see the interrelationships of your ideas because of your familiarity with your field, but someone else with another "map of the world" needs to be told how your ideas contribute to the whole picture.

Associate the unfamiliar ideas with the familiar. Relate your information directly to the needs of your audience. Use vivid imagery, contrast and comparison, hands-on experience, handouts, repetition, and—especially—visual aids to help your listeners put the main points into their long-term memory.

Be Concise

Have you ever asked a teenager if she liked a certain movie? She can go on forever telling the details, explaining everything twice, analyzing what the characters were doing, what they didn't do and why, if the leading character was or wasn't appealing, and who sat behind her in the theater. She can go through more plot twists than the screenwriter or director did when shooting the movie. But all you wanted to know was if you should go see it.

Speakers dilute their power and authority by rambling. Strive to be concise; edit your words. The acclaimed playwright Anton Chekov wrote a page-long speech for one of his characters—and then eliminated everything but "Yes." Express your ideas simply, clearly and, above all, *briefly.* Keep in mind that with the overwhelming amount of information people are receiving, they will thank you for synthesizing and selecting the most pertinent or essential data. Memorable presentations are lean and clean.

Even the experienced speaker or trainer has to adapt their content to our fast-paced world. We're used to *USA Today* with lots of pictures and simple language. Audiences have become accustomed to the accelerated images on MTV; they watch slick broadcasters, and unconsciously compare speakers to these professional presenters and their four-minute interviews. The average sound bite has shrunk from forty-five seconds in 1980 to less than ten seconds today.

Be Clear

Your audience's level of knowledge will determine the kind of material needed to clarify or prove a specific point. Socrates said a good speaker offers "to the complex souls, elaborate and harmonious discourses, and simple talks to the simple souls." If you address a graphic designer group that meets every month, their information needs may be very specific and sophisticated. A design class of community college students would be baffled by the same information.

Sometimes research engineers and scientists prepare a briefing in painful detail. In their attempts to be as exact as possible, they sacrifice clarity. If their audience can't clearly understand the material, its accuracy is of little value. This tendency to provide too much information and detail can be especially evident when staff members make recommendations to their superiors. One department head said, "This is not high school algebra. I don't want you to explain in detail how you worked out your solution. I only want your final answer."

Clarity also means choosing points that are appropriate. A vice-president of sales and marketing for a high-tech medical equipment firm had been turned down for his proposed $750,000 budget. He was denied the increase because he had come in under budget the previous year and management felt that he could do it with less in the future. Sound familiar? He came to me and said it wasn't possible for him to accomplish his goals in the coming year with less money.

We started over on his presentation. I told him to omit the details that had detracted from his main points. Instead of dwelling on why it was unfair to cut his budget, I suggested he describe: (1) which worthwhile projects would have to be eliminated due to lack of funds; (2) what could be accomplished if he kept his same budget; and (3) how his departmental goals were closely aligned with the company's goals. It worked: He got another hearing and the $750,000.

What Is Relevant?

If you were seated in the audience, what would you need or want to know? After attending one too many frothy press conferences, newspaper reporter Frank Catalano voiced the frustration he shares with many audiences who are tired of having their time wasted:

> Software companies really do need to learn how to hold a press conference. Real news conferences start with an announcement, end with questions and typically run under an hour. The goal is to have reporters leave with enough information to file a decent story, and sometimes a free cup of coffee. It benefits everyone to have news conferences that get to the point quickly and tackle hard questions. Especially when the alternative is continued gaseous "news" events in which the industry simply talks loud and long to a favorite audience—itself.

Sales representatives are finding that clients want products and services that are environmentally conscious, healthy, state-of-the-art, worthwhile, "sexy," and that give them an immediate, profitable, and gratifying return on their investment. Take another look at your audience analysis checklist in Chapter 6 and then ask yourself:

• Are these main points of *value* to your audience?

• Are they *relevant* to your audience and to your objective?

• Is this *new* information?

- Will it make the audience's job *easier*, more *productive*?

- Does it address *specific needs* and concerns?

- Will listeners be *repeating* this information to others?

- What can the audience *do* with this information?

Many of the model communicators that I interviewed said they are called on to give so many speeches that they develop a "core speech" of between three and five main points. Then they analyze their audience, write a relevant ending and beginning, and adapt their main points to that particular audience. They feel more secure about accepting speaking engagements because much of their preparation is already done.

Do your main points support your message? If not, either rewrite your message or revise the main points. Sometimes you discover significant material that should be included and, therefore, need to change your message.

Anyone can recite a laundry list of facts and data. Your audience wants *your interpretation* of those facts. What insight can you contribute to your topic? Your main points should constantly reflect your awareness of the audience. Your content demonstrates that you know who the audience is, why they have come together, and what will be valuable to them.

KEY IDEAS

- Assess what you know.

- Be selective as you gather facts.

- Contribute creative insights with intuitive analysis.

- Seek clarity rather than elaborate detail.

- Select only material relevant to your audience's needs.

Notes

1. John McDonald, "Gasoline and Clean Air," *Vital Speeches of the Day* (August 15, 1990).

Chapter 9

SELECTING SUPPORTING POINTS

"There are only two parts to a speech:
you make a statement and you prove it."
—Aristotle

OVERVIEW

Supporting points should clarify, illustrate, substantiate, and make your main points memorable. This chapter discusses how to select vivid and interesting supporting points that will help your audience relate to and visualize your ideas. If you can identify the values and beliefs of your audience, and present appropriate logical and emotional appeals that match their values, you can encourage them to accept your main points.

L. Frank Baum, the author of *The Wizard of Oz,* chose to send a lion, a tinman, and a scarecrow off in search of courage, a heart, and a brain. Two thousand years ago, Aristotle described similar traits that were necessary for an effective speaker to be persuasive. A speaker, he said, must have ethos, pathos, and logos. The *ethos*, or moral character (courage), of a person is the key factor in having an audience accept information and be persuaded. Aristotle said that we believe "good men" more fully and more readily than others. In addition, Aristotle believed that unless you could move your audience members through *pathos*, or emotional appeals (heart), it would be difficult to persuade them to change their beliefs or take action. Social psychology indicates that most people decide to act based on their emotions. They validate their actions through logic, or *logos* (brain). A speaker who uses logos is appealing to the audience's intellect through organization and facts.

Audiences can be very selfish and egocentric. They are, after all, composed of individuals, and it is said that each of us thinks about our own interests about *95 percent* of the time. A speaker can keep the listeners attentive and alert by being constantly aware of their desires and needs. Your audience analysis checklist will provide you with clues to the types of supporting evidence that will help you reach your objective.

Whether your audience is technical or nontechnical, your presentation will be weak if ethos, pathos, or logos is missing. If you were addressing a nontechnical audience about environmental concerns, you would probably rely more on establishing your credibility and providing emotional proof than you would on using a multitude of statistics. However, highly sophisticated nontechnical audiences will demand corroborating data. If you were addressing a technical audience, you would need facts, scientific analysis, and statistics to prove your points, but should also include emotional appeals.

Now that you have sharpened the focus of your presentation by a clear choice of main points, you are ready to select your supporting points. You can give evidence to support your points in the form of examples or statistics. You can offer opinions about your main points in the form of testimony from appropriate sources. You can clarify your main points by using quotations, explanations, anecdotes or comparisons, mental pictures, action, and color. You can anticipate points, identify evidence, and summarize.

STATISTICS

Technical presentations contain many statistics. Statistics reduce masses of information into generalized categories and are useful in substantiating disputable claims; but can you dramatize those numbers for better comprehension and retention? Speakers should translate difficult-to-comprehend numbers into more understandable terms, round off complicated numbers, and use statistics fairly. Who collected the data and were they objective? How current are the statistics? Too many statistics will only confuse your audience. Showing figures graphically on a whiteboard, viewgraph, slide, or computer or linking numbers to key concepts will help retention.

William D. Ruckleshaus, chairman of Browning-Ferris Industries, Inc., told me:

> The same statistics can be interpreted in vastly different ways
> by scientists and the public. For example, one out of every four
> people runs the risk of getting cancer. One out of five people,
> or 20 percent of the population, will die of cancer. When a
> scientist talks of a chemical causing ten cancers per million
> population, the average person panics. They don't understand
> that since 250,000 out of a million will get cancer anyway,
> increasing that figure by ten is not a great cause for alarm.

Your audience may lack expertise in interpreting statistics. It is your responsibility to translate your statistical data into a form that your audience can accurately comprehend.

Morley Winograd, VP of Commercial Sales of AT&T Western Region, in addressing Town Hall in Los Angeles, successfully illustrated the exponential decline of the cost of computing as a result of the increased processing power of a microchip:

> It is the equivalent of getting a Boeing 747 for the price of a pizza. Another way of looking at it is to imagine what would have happened if the same progress had been made in automotive technology: a Lexus would cost $2.00, it would travel at the speed of sound and it would go 600 miles on a thimble of gas.[1]

ANECDOTES

Anecdotes are short narratives, serious or humorous, that make a point. They support and clarify an issue, but shouldn't be expected to prove your main points by themselves. Historical vignettes or tales from other cultures can be intriguing anecdotes. Harold L. Adams, FAIA (Fellow of American Institute of Architects), RTKL Associates, Inc., pointed out that technology can serve as a trap or as a triumph, and the challenge is to determine what technology can and cannot do. He vividly illustrated this idea with an anecdote:

> When the art of writing on papyrus was first developed, an example of this new technology was rushed to the king. It was explained that now he no longer had to rely on memory and the kingdom could retain a history of what had gone on before. The king looked at the example with sorrow and said, "No, from now on, everyone will forget." The king couldn't anticipate that two thousand years later, the world—and you and I—would be suffocating under mounds of paper. And he was right, people would forget.[2]

Audiences enjoy and retell anecdotes and war stories. Start collecting and compiling them long before you need to use them. Unique anecdotes can be found daily in newspapers, magazines, and conversations.

EXAMPLES

Your audience will not be persuaded if you make too many abstract generalizations and don't support these statements with statistics, facts, and examples. In the audience analysis checklist, I ask my technical clients to predict if the audience will be friendly, hostile, or have a "show me" attitude. Examples are an excellent way to satisfy this attitude. *If you make a general statement, give an example.* Examples illustrate a concept, condition, or circumstance. Use descriptive adjectives that will narrow down a generalization and help your audience form a clear mental picture.

There are two kinds of examples. A factual example describes an actual situation in detail. This type of example is highly persuasive. Kathryn Kelly, an environmental toxicologist, was brought into a small community of 400 people in Alaska that wanted to block the building of a garbage incinerator because they were concerned about dioxin emissions. She was asked to do a study, assess the pros and cons, and report to the townspeople. As she went through the town, she noted the smoke emanating from woodstoves in their homes and she began to gather data about this heat source.

The risk of dioxin is usually compared to the risk of developing cancer from smoking, the pH's in charbroiled steaks, or aflatoxin in peanut butter. Kelly didn't feel the townspeople would relate to these examples. In her presentation, she acknowledged their concerns about the emissions. She went on to point out that the 400 woodstoves in the town were emitting dioxins equivalent to 1200 incinerators, and they only wanted to build one. By using terms familiar to the townspeople to illustrate the actual risk, Kelly's example reduced their apprehension.

Hypothetical examples also provide a valuable means of clarifying an idea. They involve the imagination of the audience, but are not as powerful as factual examples. This type of example involves the "what if" scenario. Jack Boyd, Executive Assistant at Canon Company, Inc., spoke to the Virginia Works Conference. He gave one negative hypothetical scenario of the year 2026 and what might happen if the U.S. failed to alter its educational system. Then he gave another scenario:

> Let's listen to Professor Samuel's class. This time our professor is teaching five hundred thousand students who are scattered across many states. His presentation is transmitted via an on-line telecommunications system, either for real time reception or for digital storage for later retrieval. He is employed by Middle Atlantic University, a major learning institution that grew out of a consortium of regional colleges and universities. In schools nationwide, children are learning pure academics alongside work skills, scientific and mathematics theory coupled with practical application, and long term learning values coupled with key behaviors valued by employers. Ladies and Gentlemen, is this a realistic image of the future?[3]

This vivid example painted a striking picture of the need for creating a strong workforce for the future.

Use examples that are representative and not deviations from the rule. Your audience may question an isolated incident. How many examples should you mention? One example may not be enough, and although you may give one example in detail, it may be best to use one or two others for your listeners' reference so they can see patterns. If your topic is controversial and your audience is thinking of all the negative examples that refute your example, then one example will not carry enough weight. Avoid sprinkling your presentation with too many details and examples without providing enough general statements that tie them together.

When giving examples, think globally, as your audience will likely represent international companies. Support and enhance your image by using case studies or client references from China, Brazil, or other worldwide connections. If referring to yourself or an industry, use "North America" instead of "United States."

EXPLANATIONS

Explanations are usually simple expositions or descriptions that serve to make a term, concept, process, or proposal clear and intelligible. They are often reinforced by examples, statistics, or other forms of supporting evidence. Explanations also tell your audiences how something came to be, how something happened, or how something is done. University of Illinois professor David F. Linowes used this cause-and-effect type of reasoning in a speech to the White House Conference on Libraries and Information Services. He explained how the capability of some nations to collect, store, manipulate, and disseminate information

> has caused a greater disparity among developing nations and
> sophisticated nations than have the differences in material
> wealth. This flood of new knowledge makes it impossible for
> them to catch up and become self-sufficient, resulting in a new
> and more sinister form of colonialism. This perspective embitters
> relationships between us and the have-not nations and will
> increase in the future.[4]

RESTATEMENTS

A restatement is a reiteration of an idea in different words or in a different way. It has a subconscious persuasive impact and can make a point memorable. Be clear and precise and avoid repetition that is boring.

Restatement was used well by Paul M. Weyrich, president of the Free Congress Research and Education Foundation, in an address to The Washington, D.C. University Club. Weyrich pointed out, "The computer cannot solve all of our problems for we must first define what our problems truly are." Then he used poetry to repeat the thought in a different way. "In 1928," Weyrich said, "before the computer age had dawned, American poet Archibald MacLeish wrote the following lines: 'We have learned the answers, all the answers: It is the question that we do not know.'"

And he reiterated his thought again using the following colorful analogy:

> Yes, we do have the answers, lots of answers, more answers than
> we know what to do with. Our computers are crammed with
> answers. But what is the question that will endow those random
> facts with significance, and purpose? Like orphaned keys found

in an attic drawer, facts by themselves are useless, however
bright and shiny they may seem. Better to have a lock without a
key, a puzzle in steel to solve, than keys to nothing.[5]

Several times in my classes, I emphasize the need to edit your words, since
20 percent of what you say makes 80 percent of the impact. I give examples of
this idea and illustrate it with a visual. Several former students have told me they
remember the 20/80 rule from my class and are motivated to edit their words in
their communications. Repetition made the concept memorable years later.
Telling, rephrasing, and expanding on an important idea will give your audience
time to process the information and put it into long-term memory.

QUOTATIONS AND TESTIMONY

Prime Minister Winston Churchill studied quotations intently: "Quotations, when
engraved upon the memory, give you good thoughts. They also make you
anxious to read the author and look for more." Quotations can be a way of
saying that you and a well-known person such as Einstein, the pope, or a Nobel
Prize-winner think alike. The statement doesn't have to come from an expert if
the person says precisely what you mean. Children, friends, parents, or the store
clerk can be great sources. You can find hundreds of books full of quotations on
every topic imaginable, but it is more important to start keeping your own
notebook of favorite quotations.

Testimony refers to the opinions or conclusions expressed by others and can
be used to support your points. Testimony is an opinion about facts and, therefore, is not as effective as the facts themselves. The person quoted should be
qualified by training or experience as an authority respected by your audience
and not be unduly biased. An environmental engineer who cites only other
environmental engineers will not convince an audience who thinks the speaker's
company is out to ravage the environment. Sometimes, reluctant testimony can
be very influential. A highly respected person who has always advocated one
position and then admits to a change of mind can convince others to examine
their beliefs.

This is a list of favorite quotations that I have collected and found relevant to
my scientific, engineering, and technological clients.

Great Thoughts to Enlighten, Encourage, Persuade, and Inspire

*"The man who can think and does not know how to express what he thinks is at
the level of him who cannot think."*

—Pericles

*"Knowledge is of two kinds. We know a subject, or we know where we can find
information upon it."*

—Samuel Johnson

"The dangers that face the world, can, everyone of them, be traced back to science. The salvations that may save the world will, every one of them, be traced back to science."

—Isaac Asimov

"The main obstacle to progress is not ignorance, but the illusion of knowledge."

—Daniel Boorstin

"Either he succeeds by offering to the reader only superficial aspects, thus deceiving the reader by arousing in him the deceptive illusion of comprehension; or else he gives an expert account of the problem, but in such a fashion the untrained reader becomes discouraged from reading any further."

—Albert Einstein

"Good design is intelligence made visible."

—Frank Pick

"Technological wizardry is not an end in itself, it is desirable only if it makes for human welfare, and this is the test that any tool ought to be made to pass."

—Arnold Toynbee

"We owe a lot to Thomas Edison—if it wasn't for him, we'd be watching television by candlelight."

—Milton Berle

"Whenever science makes a discovery, the devil grabs it while the angels are debating the best way to use it."

—Alan Valentine

"The universe is full of magical things patiently waiting for our wits to grow sharper."

—Eden Philpotts

"Science is facts; just as houses are made of stones, so is science made of facts; but a pile of stones is not a house, and a collection of facts is not necessarily science."

—Jules Henri Poincaré

"I can give you anything you want, but time."

—Napoleon

"No one would have ever crossed the ocean if he could have gotten off the ship in a storm."

—Charles Kettering

"The secret of managing is to keep the five guys who hate you away from the guys who are undecided."

—Casey Stengel

"The gift of fantasy has meant more to me than my talent for absorbing positive knowledge."

—Albert Einstein

"A thing to be simple needs only to be true to itself."

—Frank Lloyd Wright

"Most people are other people. Their thoughts are someone else's opinions, their life a mimicry, their passions quotations."

—Oscar Wilde

"You can't say that civilization don't advance, for in every war they kill you a new way."

—Will Rogers

"He had nothing to say, and he said it."

—Unknown

"If the primary aim of a captain were to preserve his ship, he would keep it in port forever."

—St. Thomas Aquinas

"Be yourself. Who else is better qualified?"

—Frank Goblin II

"If I have ever made any valuable discoveries, it has been owning more to patient attention than to any other talent."

—Sir Isaac Newton

"If you obey all the rules, you miss all the fun."

—Katherine Hepburn

"Eloquence is the art of saying things in such a way that those to whom we speak may listen to them with pleasure."

—Blaise Pascal

"In groups of chimpanzees, the leaders are not necessarily the strongest or the fiercest, but the ones who make the most friends."

—William F. Allman

"We shall not cease from exploration
And the end of all our exploring
Will be to arrive where we started
And know the place for the first time."

—T. S. Eliot

EMOTIONAL APPEALS

"The world has kept sentimentalities simply because they are the most practical things in the world. They alone make men do things. The world does not encourage a perfectly rational lover, simply because a perfectly rational lover would never get married. The world does not encourage a perfectly rational army, because a perfectly rational army would run away."

—Gilbert K. Chesterton

As you choose your anecdotes, statistics, testimonies, and other supporting materials, keep the emotional needs of your audience in mind. Model communicators in technical and scientific fields say that they do not hesitate to use emotional appeals in technical presentations. They are able to break up the constant flow of dry, complex information by using examples, anecdotes, explanations, and quotations. They are aware emotional appeals can be very seductive and persuasive. These supporting points give variety to the pace of the presentation, and make it easier and more comfortable for the audience to follow the content. Good speakers, regardless of their subject matter, choose supporting points that are vivid and always directed toward the immediate needs of the audience.

Abraham Maslow's hierarchy of needs lists five levels:

1. *First level*—the basic needs of survival and sex;

2. *Second level*—the needs of safety and security;

3. *Third level*—the need of belonging;

4. *Fourth level*—the needs of self-esteem, recognition, and competence;

5. *Fifth level*—the needs of self-actualization, challenge, and realization of potential.

A speaker must determine at what level the needs of the audience are unsatisfied, deal with those concerns, and only then advance to appeals at the next level.

A manager from an engineering firm was asked to talk to a high school class about achievement and leadership, which are qualities in the fourth level of Maslow's hierarchy of needs. He was savvy enough to recognize that most of the high school students were primarily concerned with the third level of belonging and being accepted by their peer group. Since all these needs were not being met, only the popular students or sports stars would be interested in the next level of leadership. So he cited examples of the football team, the drama club, and the school newspaper needing teamwork. He pointed out that a good leader was required to make each of these teams successful. The students could relate to these examples and listened to his talk because they wanted their school to be number one.

The best way to get someone to do something *you* want them to do is to get *them* to want to do it. People do what they do based on *their* values, not on yours. Why would this be important to them? Emotional appeals can be *positive* or *negative*. You can tell them they will *forfeit* peace, safety, or harmony unless they act in a certain way or they will *obtain* peace, safety, or harmony if they do act in a certain way.

My daughter told me that her friend who has a brain tumor surveyed hospitals to find the best brain surgeon. She narrowed down the field to three doctors and based her final decision on the one with the best "bedside manner." It seems totally illogical to me to be going under anesthesia for an extremely complicated operation and feeling reassured because the neurosurgeon will be warm and comforting when, and if, she wakes up. But even in this extreme case, emotional appeal was more important than logical reasoning.

Consider some of the following emotional appeals. How can you help your audiences to:

- Feel better about themselves?

- Avoid being boxed into a corner?

- Make their work easier, not harder?

- Be thought of as honest, fair, kind, and responsible?

- Finish this proposal, update, or negotiation, and move on to something else?

- Know the truth?

- Feel that they are doing something that matters?

- Avoid failure and preclude future risks and trouble?

- Meet personal goals without violating their integrity?

- Be listened to?

- Avoid surprises and abrupt changes?

- Be liked?

- Be challenged?

- Gain power?

A computer software sales rep told me that initially his clients go through extensive logical analysis to determine if his product will meet their needs. Price is usually *not* the determining factor. The final decision is a "gut level" emotional one based on many of the factors in the preceding list.

This sales rep recognizes his clients' fear of failure. They want to make a safe decision and avoid future trouble and risks. To answer this need, he stresses the successes that other clients with similar job responsibilities have had with his firm's products. Another emotional appeal he uses is that purchasing his product will gain approval from management. His product is proven and has a track record of success.

He taps into the fifth level of Maslow's hierarchy of needs when closing a sale. He appeals to the clients' desire to achieve their potential. He compliments them on making an excellent decision. They have addressed a problem and answered the challenge in a competent, thorough way. The clients are left feeling good about themselves and their abilities.

Realize that the justifications that people give you for their decisions are rarely the real reasons a person disagrees with you. Most decisions are based on *fear,* and if you can discover that insecurity and resolve it in your presentation, you are more apt to get approval.

Commitment

Your emotional appeals will be accepted in relationship to your perceived credibility. Billy Graham can talk about hell and damnation and be very emotional because his credibility is extremely high with most of his audiences. A renowned scientist or engineer in his profession for thirty years can freely use emotional appeals. But if you are unknown and inexperienced, you need to use persuasive facts and statistics to support your points before you add many emotional appeals.

If you make an emotional appeal, your commitment should be heard in your words and voice. It is disconcerting to have presenters using emotional appeals who seem to have distanced themselves from the whole situation.

If you offer a solution, describe it in vivid terms and paint a colorful picture of it. Recently I was told of a gentleman who came to the Northwest to raise funds for a project. He intended to create an island in the Bahamas and build a luxurious gambling community on it. No such island existed, except in his imagination. He had an architect draw renderings and create a three-dimensional model of the proposed island and gambling resort. This gentleman was so convincing and painted such a clear, vivid picture that people were jumping on the bandwagon to invest money in this dream. A prominent psychologist I spoke with said that, although he had reservations, he was prepared to give the businessman a check when the entrepreneur became ill and died. The clarity and emotional appeal of this promoter's vision was hard to resist.

When you are gathering your supporting materials, remember the lion's, the tin woodsman's, and the scarecrow's quest for courage, a heart, and a brain from the *Wizard of Oz*. You must be prepared to confidently supply both logical and emotional arguments in your presentation. All three elements must be present for you to successfully persuade your audience to accept your ideas.

KEY IDEAS

- Bring your main points into sharp focus with convincing supporting points.

- Use a variety of types of supporting points such as statistics, examples, testimony, and quotations throughout your presentation.

- Start compiling your collection of quotes, interesting facts, and anecdotes for future presentations.

- Be logical, but also address emotional needs.

- Review your audience analysis to choose appropriate logical and emotional appeals.

Notes

1. Morley Winograd, "Social Contract for the Information Age," *Vital Speeches of the Day* (November 1, 1996).
2. Harold L. Adams, "Technology—Trap or Triumph?," *Vital Speeches of the Day* (November 15, 1987).
3. Jack Boyd, "Creating Yesterday's Workforce. The View from 2026," *Vital Speeches of the Day* (May 1, 1997).
4. David F. Linowes, "The Information Age," *Vital Speeches of the Day* (January 1, 1990).
5. Paul M. Weyrich, "A Conservative Vision for America's Future," *Vital Speeches of the Day* (October 1, 1990).

Chapter 10

ORGANIZING YOUR CONTENT

*"A forest of facts unordered by concepts and constructive relations
may be cherished for its existential appeal, its vividness,
its pleasure, or its nausea, yet it is meaningless, insignificant,
and usually uninteresting unless it is organized by reason."*
—Henry Margenau

OVERVIEW

*Your listeners need to be able to relate the facts and data you present to a frame
of reference. A strong organizational pattern will provide that structure and
bring order and clarity to your presentation. This chapter discusses several
patterns that will enable your audience to follow your reasoning and reach the
same conclusion that you do. Structure can serve as a memory device for you to
anchor your thoughts. This chapter also emphasizes the need for transitional
words and phrases to signal important relationships between key points.*

You have defined your objective, analyzed your audience, and gathered your
main points and supporting points. You have decided on your conclusion and
beginning. Now, how do you arrange the key points in the body of your presentation in an order that will effectively lead your audience by easy stages toward
acceptance of your ideas? If you are going to London from New York, would it
be better to travel by hot air balloon, jet plane, cruise ship, or sailboat? Your
answer depends not on your destination but on the purpose of your trip, your
objective. Your objective may be to get there as quickly as possible, to have a
leisurely trip, to travel in the least expensive manner, or to visit all the tourist
attractions in between. You select the way to go according to what you want to

accomplish. All of the transportation methods mentioned will get you to London, but not all will necessarily accomplish your objective.

In a presentation, your choice of organizational pattern should be based on your *subject matter;* the *needs, values,* and *knowledge* of your audience; and your *perceived credibility* with the audience. Whatever the organizational pattern you choose, it should clarify your message, lend interest to your topic, emphasize those points that you want your audience to focus on, and exhibit your creativity and grasp of the topic.

The president of the local International Association of Business Communicators gave a presentation at the group's monthly meeting. Her objective was to inform newcomers about the association and encourage people to join. Her message was that membership in the IABC could be a valuable career boost. She began by saying, "I could give you the usual facts about our organization, but I think you can get a better picture if I tell you what it has meant to me." She outlined her career history and described how the IABC had been an important factor in each of her career moves. The way in which she arranged the information showed the connection between IABC membership and career advancement. Instead of giving the audience a dry list of salient points, she engaged the listeners with a personal story and won them over.

IMPOSING ORDER ON CHAOS

Why be organized? By developing the structure of your presentation, you and your audience will be able to remember key ideas. You lighten the strain on the listeners so they will be able to immediately grasp the relationship of your points and the meaning or relevance of your ideas. In addition, you will be more credible. Studies show that audiences perceive very quickly when a speaker is organized, and based on that perception, grant a degree of credibility to a speaker. Organization also provides you with a scheme that will propel you forward and give your presentation a feeling of momentum.

GOOD NEWS, BAD NEWS

If you need to present negative information, the audience probably will not listen closely until you deal with the controversial topic. One speaker used humor to lessen the impact of a negative quarterly report. He began:

> I have some good news to relate from research and development, but first I want to deal with the financial setback this last quarter. I don't believe in the rationale of a scientist who had been unjustly accused and convicted of treason in a mid-eastern country and found himself forgotten in a prison in the midst of the desert. He was given a cellmate who tried to convince him to make an escape attempt, but he refused. The cellmate went ahead with his plans on his own and made his escape. However, he ran out of food and water and nearly succumbed to the intense heat. Despondently he returned to the prison and related

his traumatic ordeal to the scientist, who astonished him by agreeing, "Yes, it was horrible. When I tried it, I also failed." The cellmate indignantly demanded, "For heaven's sake, man, when you knew I was going to make a break for it, why didn't you tell me what it was like out there?" The scientist shrugged his shoulders and replied, "Who publishes negative results?"

The speaker's approach relaxed the audience and they knew they would be getting the full story on the status of the company.

Abraham Lincoln remarked, "My way of opening and winning an argument is first to find common ground or agreement." Establishing commonality, an agreement of ideas in the beginning, will lead to rapport and trust. If you start to threaten deep-seated values and strongly held beliefs, your audience will not listen to your persuasive argument, but will start thinking of why you are wrong. A thorough audience analysis should uncover those beliefs.

Many of my clients ask if they should present an opposing view. A speaker addressing an educated and intelligent issue-oriented audience will be more effective if she presents both sides of any argument and meets the principal opposition head-on. If your audience is hostile to your view, it is also preferable to present both sides, so you look fair. When an audience is in agreement with you, almost anything you say will reinforce beliefs. The best evidence points to the general principle that if the audience agrees with you and knows little of the opposing position, stronger temporary effects will be produced by showing only your side of the controversy. However, in most situations it is better to meet the principal opposition arguments. Your listeners will read about them or hear about them in any case, and if you expect to be persuasive in the long run, you may choose to deal openly and frankly with the material contrary to your message. You need to decide this on a case-by-case basis.

How do you reestablish your credibility and trust when presenting to key decision makers who are aware of "skeletons in the closet?" Realize past shortcomings will be *foremost in the minds of your audience and affect their perception.* By owning up to mistakes and focusing on solutions, you can turn potential liabilities into assets. For example, one engineering firm had many personnel and project management problems on a government contract. Their reputation was sullied by the cost overruns and delays, plus animosity built up with the government agency. Four years later, the same site manager and his staff were making an oral proposal for a new project before a government committee. Three of the five committee members had been involved with the previous fiasco. What to do?

I suggested they candidly address the past problems in the first ten minutes of their presentation. To avoid improvisation, I had them memorize their statements. "You are well aware of the previous problems. The project manager and senior engineer who worked on that contract are no longer with the firm. We will describe some of the *controls* that we have implemented so that this will never happen again. We have brought *letters of reference* from several of our recent clients attesting to the *successful completion* of jobs *within or below budget* and *on time*."

If you have high credibility, your listeners may be more patient with the way you parcel out good or bad news. If you do not have seniority or authority, they will probably demand you address their concerns immediately.

ORGANIZATIONAL PATTERNS

> *"Normal sequence of product design:*
> *1. Management announces the product.*
> *2. Technical writing publishes the manual.*
> *3. Engineering begins designing it."*
>
> —William Horton
> *Horton's Law*

A presentation is often promised and publicized long before the speaker has any specific idea of what he or she is going to say. The presenter looks at the now unfamiliar abstract/outline that was hastily submitted months ago, and winces. "I can't emphasize that! Our research is going in a different direction," or "Our product focus has shifted," or "We have reconfigured that process." You will need to creatively reconcile the old outline with your up-to-date perspective.

Regardless of the circumstances, ask yourself which plan will take into consideration the different levels of knowledge within your audience, their most pressing needs, the hidden agendas? Which plan will illuminate your subject? Which plan will automatically feed into a persuasive finale? Your finish should have a major influence on your choice of pattern. Here are some examples to consider.

Chronological Order

This pattern organizes your ideas by time such as the past, present, and future. It has a built-in sense of forward movement. Your topic may dictate this organizational pattern. In a scientific or technical presentation, for example, speakers may need to describe the steps in an assembly process of an operational cycle. It is necessary to present complex, sequential instructions in the proper order and not leave out critical steps or instructions.

Rather than give a dull minute-by-minute or year-by-year accounting, try to cover periods of time and make them more vivid. For example, your main points could be that sales were slow from 1993 to 1995, peaked during 1996 to 1998, and held steady until the present. Robert M. Price, President of PSD, Inc., imaginatively discussed generations of computers by their role in society rather than by technological progression. He said a societal view of computers showed three generations, each one roughly equivalent to a twenty-year human generation. He compared the first computers with the European immigrants that came

---

ORGANIZATIONAL PATTERNS

Chronological Order
(Time)
Past
Present
Future

Topical Order
Social
Political
Economical
Cultural

Escalating Pattern
Local
Regional
National
Global

SAFW (Say a few words)
Statement
Amplify
Few Examples
Windup

Experiential Order
Disbelief
Reexamine Evidence
Belief

Pro and Con
All the Points in Favor
All the Points Against
Conclusion—Which Side is Best?

Cause to Effect

Scientific Method
Background/Purpose
Materials/Methodology
Results
Recommendations

Problem Solution
Attention
Problem
Solution
Visualization
Action

---

to America. Then he took his audience through the second and third generations, and said of the third generation, "They will be better educated and more affluent than their predecessors, based on a spectrum of technology their grandparents couldn't imagine. They will be literate, articulate, and completely integrated with their human partners."

Topical Order

Discover interesting ways to divide your subject into different parts. It is quite common for project managers to divide their progress reports into categories of *personnel, equipment, time*, and *budget*. Rick Chappell, associate director of science for NASA Marshall Space Flight Center, divided a presentation about proposed cooperation between the U.S. and Russia joint venture in exploring Mars into its *social, political, economical*, and *cultural* aspects. Topical divisions shouldn't be a random list of items. The divisions used should relate to one another and be of approximately equal importance. If you are presenting four results of research, ask yourself:

- What order would be most *logical* and natural considering your audience's level of knowledge?

- What information would be most *valuable* to them?

- What would be the easiest way for them to *understand* the information?

Take into consideration the memory curve that exists during any presentation. Attention will usually be highest in the beginning and toward the end. Your most important division should be positioned up front and reviewed toward the end. Maintain a sense of forward movement to your irresistible conclusion.

Escalating Pattern

Speakers who use this pattern can organize their material *smallest to largest, easiest to hardest*, or *inexpensive to costly*. A pilot told me that it was critical in his training program to learn the first level of information and practice it before he progressed to the next level. Similarly, if you are teaching a person a new software program, you must progress through its applications from the most *basic* to the more *complex*. Each application must be learned before advancing to the next one. An overload of information at the beginning is useless.

This is a pattern frequently used at annual sales meetings. For example, a hospital equipment firm presented strategy to distribute their products on a *local, regional, national,* and *international* level. This progression was combined with a chronological analysis of *past* accounts, *present* sales figures, and *future* outlook.

SAFW

How many times have you been in a meeting and the chairperson has asked you to "say a few words" about the topic under discussion? SAFW stands for:

- <u>S</u>tatement

- <u>A</u>mplification

- <u>F</u>ew examples

- <u>W</u>indup.

This formula will help you organize your thoughts so that you sound intelligent and poised. Begin with a strong *statement* that encapsulates your feelings, insight, and belief about the subject. Then *amplify* that statement. Transitional phrases could begin with "that means," "therefore," "however," or "for that reason." Cite one or two *examples*, depending on your time constraints, and *wind up* by making a recommendation or offering a solution and circling back to your original statement. Try using the SAFW formula the next time someone asks you a question.

Experiential Order

Experiential order reveals how we came to an understanding or *developed a belief*. This arrangement of ideas can be very effective if you are talking about a certain methodology or experiment that you disbelieved and now agree with. For example, a person might tell how he doubted citizen garbage recycling efforts would work. He reports that after re-examining the evidence, and personally participating in recycling, he now believes in allocating more funds to support recycling projects. If the speaker's credentials are highly regarded, this can be a convincing way to organize material.

Pro and Con

This pattern gives all the *points in favor*, all the *points against,* and then concludes *which side is best*. You may also give the *advantages* or *disadvantages* of two sides of an issue and end up offering a compromise. If you are talking to your peers, you may be able to jump into the heart of a technical discussion on advantages and disadvantages. However, general audiences may require more background material to bring them up to a basic understanding of your main points.

Cause to Effect

This pattern points out that *one thing can lead to another*. Of course, you have to prove your thesis is true. Many presenters will find a cause-effect organizational pattern ideal for discussing environmental concerns, occupational hazards, or results of research.

One presenter was trying to convince his audience of the dangers of mercury-amalgam dental fillings. *To prove that mercury was harmful*, he cited the classic example involving the felt hatters in the nineteenth century. At that time, felt hats were dipped in a mercuric-nitrate solution to make the felt easier to shape. The workers were subject to severe mercury poisoning after inhaling mercury vapors and absorbing mercury through the skin. This caused tremors, incoherent speech, muscular dysfunction, and even feeble-mindedness. The author Lewis Carroll portrayed these effects with his memorable character, the Mad Hatter in *Alice in Wonderland*.

The presenter went on to describe how current research had turned up many *patients with various health problems* who also had *mercury in their dental fillings*. When the fillings were removed, the patients reported amazing relief. In this case, as with any presentation using a cause-effect organizational pattern, the audience had to decide whether the causal relationship was effectively demonstrated.

Scientific Method

This organizational pattern gives the *background of the problem* and *purpose of the research,* lists the *materials used,* explains the *experimental methodology,* gives *results,* enumerates *implications,* or makes *recommendations.*

Laboratory scientists and technicians at the bio-tech company Immunex are asked to present scientific data at monthly project meetings. These twenty- to thirty-minute oral presentations are before an audience with a wide range of scientific expertise and backgrounds. Phil Morrisey, Director of Molecular Immunology at Immunex, elaborated on the expectations for these project updates:

- Briefly explain why the experiment(s) was/were done in the context of project goals.

- Describe methodology.

- Present the experimental data in a clear and concise manner.

- Utilize graphs, charts, and tables that are easily understood.

- Provide an intelligent interpretation/critique of the data.

- Indicate what further experiments need to be done (when applicable).

- Participate in discussion of the data.

David Suzuki, in his book *Inventing the Future,* questions the rote, passionless way in which students learn to communicate their research. He tells of screening more than a quarter million flies carrying chromosomes exposed to powerful chemical mutagens and recovering a temperature-sensitive paralytic mutation:

> At 22° Celsius, the flies could fly, walk, mate—they seemed normal—but when shifted to 29°C, they instantly fell down, completely paralyzed. When they were placed in a container kept at 22°C, they were flying again before they hit the bottom! It was a spectacular mutant, and, when we found it we screamed and danced and celebrated....
>
> And how did we write up our results? We riffled through all our records, selected the ones that said what we wanted and then wrote the experiment up in the proper way: purpose, methods and materials, results and so forth... in a way that suggested we began with a question and proceeded to find the answer. That's because the report was a way of "making sense" of our discovery, of putting our results into a context and communicating it so colleagues could understand and repeat...what we did. But it conveyed nothing of the excitement, hard work, frustration, disappointment and exhilaration of the search—or the original reason we started the search for paralytic mutants![1]

Problem Solution

Many of your presentations will lend themselves to a problem-solution organizational pattern. Sales presentations, including TV commercials, often use this pattern. First, focus the *attention* of the audience on a particular *problem*, sometimes making the audience aware there *is* a problem. The next step is offering a *solution* or satisfying a need. Then take your audience through the *visualization* step, or what will happen by implementing the solution, and finally ask for *action*.

One of my engineering clients submitted a written proposal for a bid on a large EPA (Environmental Protection Agency) contract. They were one of four firms selected to create and deliver a response scenario for the investigation of the cleanup of a mining site. The presenter developed a forty-five-minute *problem-solution* presentation. He focused the *attention* on the site graphically drawn with the detailed mining site, buildings, and a town one-half mile down the stream which eventually led to a river. He explained it would be the objective of his firm to do an eighteen-month investigation; deal with immediate threats to humans, animals, and fish; and complete as much remedial work as possible. He showed that his team clearly understood all the problems. He listed the *problems*/assumptions such as iron tailings, sediment, runoffs, and possible contamination that were shown on the conceptual model. Then he offered *solutions* clearly diagrammed on another detailed chart. He proposed program management and Web sites that would facilitate communications with town meetings involving shareholders. He asked them to *visualize* the time and work that would be saved and how costs would be reduced. The team finished by asking for *action* or the contract. They reviewed the experience and positive performance scores of their company as final proof of how this contract would be carried out efficiently and effectively.

EXPERIMENT TO FIND THE BEST PATTERN

These are a few of the different patterns you can use. The most efficient and appropriate pattern may not be apparent at first glance. Outline and condense your information into six minutes and experiment by putting this information into different organizational patterns. By trial and elimination, you will discover that one pattern seems to simplify and illuminate the material better than the others. As you restructure the ideas, the different arrangement may also suggest new ways to clarify your information. You may find that you lack evidence in one part or another, or that your supporting points are not of equal weight. You might choose to reverse the order of main points, because your audience needs to be aware of specific information before they can grasp your last point.

Your organizational pattern should be flexible, so that you can adapt to unexpected audience reactions or situations. Just as you may have to make a long detour on your journey because of a major storm, you may have to spend extra time explaining a concept, and then condense another part of your speech to stay within your scheduled time limit.

Many speakers hesitate to preview or review points, because they think the reiteration will be boring. However, audiences usually welcome the repetition if the point is important. References to ideas previously made should relate to the single purpose the speaker hopes to achieve. Several model communicators have told me that they follow the old adage:

• Tell them what you're going to tell them

• Tell them

• Tell them what you've told them.

Previewing an agenda, discussing the material, and then summarizing will help your audience to remember essential information.

Model communicators emphasize that most listeners appreciate smooth transitions that tie sections of a speech together with a ribbon of relevancy. Use words and phrases to help the listener connect one point with the next and see their relationship to the whole message. Verbal signals will help even the distracted audience member to eventually catch up.

You can ask listeners to shift from the general to the specific by saying "for example" or "one instance." Alert them to a change of direction and guide your audience along by saying, "nevertheless," "but," or "on the other hand." Summarize what you have said and forecast what is coming up by saying, "We have looked at the recent research on termites and now we are going to examine how the principles of this experiment relate to our present insect population."

If you have a clear organizational pattern, the audience can mentally check off points as you make them. If you are presenting an oral proposal to a government organization that has requested certain criteria, they may be physically checking them off a required list. Don't jump abruptly from one point to another. You may have everything clearly worked out in your mind, but the audience needs to be prompted so they can discover the continuity within your presentation.

In the same way, clearly signal to your audience the interrelationships among your points. When you need to establish a sequence, use such terms as "subsequently," "after," "next," or "now we come to the third step in the process." Connect statistics with something you said earlier and relate each section of your presentation to your overall thesis. Use such phrases as "another viewpoint," "and yet," "a final relationship," "however," or "it is equally important." If you wish to report implications of research, you could use phrases such as "in the future," "we predict," or "the outlook in the years ahead." Look back over your presentation to see where you might insert connecting statements that not only help the audience keep up with you, but also lead toward the acceptance of your ideas.

KEEP EVERYONE—INCLUDING YOURSELF—ON TRACK

Model communicators will choose an organizational pattern that emphasizes certain points and de-emphasizes others. Most audiences that have to absorb and remember scientific and technical information appreciate a road map. They're most comfortable with speakers who tell them what's coming: "I would like to discuss how this cleanup affects the environment locally, state-wide, and nationally" or "I would like to state the results and then give my recommendations for future experiments."

A clear design will help you, the speaker, remember the order of your points, even if you should lose your place. One of my clients struggled with her presentation. She had a long list of critical points, but there didn't seem to be "rhyme or reason" to the sequence of ideas. I suggested that she group the ideas under *Past, Present,* and *Future.* Immediately there was a natural and progressive flow to the information and she didn't have to refer to her notes.

Don't make the mistake of thinking that elaborate visuals can substitute for a strong structure. A road map not only makes your audience feel comfortable and secure, but also reminds you, as a leader, of your direction.

One of my clients was asked to speak to two different audiences on the same topic. He was given thirty minutes for his first presentation and two hours for the second one. When he told me he would cut his two-hour speech to fit into thirty minutes, I cautioned him against this method. I told him each presentation and audience demands a different strategy. He ended up using a topical approach for the two-hour speech and a simple problem-solution approach for the shorter presentation.

Each issue of the British magazine *Structural Engineer* carries the following quote on its masthead:

> Structural engineering is the science and art of designing and
> making, with economy and elegance, buildings, bridges, frame-
> works, and other similar structures so that they can safely resist
> the forces to which they may be subjected.

If you can give your speech an economical and elegant structure, you will find that it, too, can withstand unforeseen circumstances. You will also find that you look at your material in a fresh, new way. Examine the logical sequence of your material and judge its probable effectiveness as a whole. It doesn't matter which pattern or combination of patterns you choose, as long as the information is presented in a way that promotes understanding.

KEY IDEAS

- Choose an appropriate organizational pattern that clarifies your message and helps the audience remember it.

- Experiment with different patterns to find the best fit.

- Practice using formulas (SAFW) to organize your ideas when you answer questions or make brief comments.

- Create a feeling of forward movement in your presentation.

- Move your audience from point to point with transitional phrases and highlight the interrelationships of your ideas.

Notes

1. David Suzuki, *Inventing the Future* (Toronto: Stoddart, 1989). Used with permission of the author.

Chapter 11

CREATING VISUALS—FROM LOW TECH TO HIGH TECH

"The task of the designer is to give visual access to the subtle and the difficult—that is, the revelation of the complex."
—Edward Tufte
The Visual Display of Quantitative Information

OVERVIEW

Effective visuals grab an audience's attention and quickly convey abstract concepts and complicated relationships. They increase the audience's retention of your material, clarify and support your points, and play a major role persuading your listeners. This chapter provides you with a step-by-step approach to the use of visuals and electronic presentations. It discusses advantages and disadvantages of various types of visual media. Technology is making it easier to create dazzling effects, but visuals should not upstage or replace you or your message. Even if the equipment fails, your message should remain intact.

Model communicators have spent time and effort to master new technology tools and incorporate them into their presentations. Whether you use low tech or high tech, you can become more effective in creating visuals and more comfortable in using them.

Recently, at the end of a three-day technical conference, I couldn't absorb another particle of information and was eager to make my departure to the airport. But I hesitated when I noticed a presentation at 5 p.m. on digital imaging. It didn't look very promising when I glanced in the room and saw only thirteen people scattered about. James Lambert, Technical Group Leader of Biomedical and Visualization Systems at Jet Propulsion Laboratory, was introduced and began talking to the group about noninvasive 3-D imagery as an

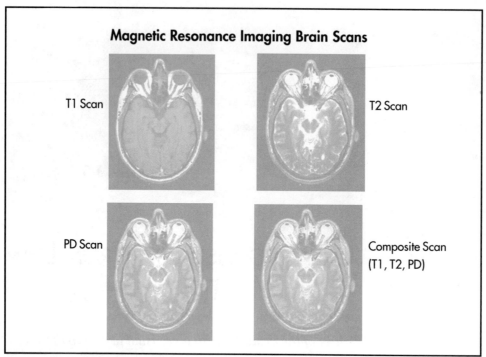

Magnetic Resonance Imaging Brain Scans

T1 Scan

T2 Scan

PD Scan

Composite Scan
(T1, T2, PD)

Courtesy of James Lambert[1]

efficient tool in medical diagnosis. In spite of the low attendance, he shared his research with genuine excitement. He used a series of colored viewgraphs to simultaneously show four slides. Each slide represented an axial section of the brain taken with three different MRI (Magnetic Resonance Imaging) scans—T1, T2, proton density, and the fourth slide was a color composite of the first three slides. Lambert discussed how this imagery helped gauge the efficacy of drugs to alleviate neurological disorders, such as Alzheimer's disease. Then he showed a video with animated 3-D renderings, formed from the sections of the brain that we had seen on the viewgraphs. Lambert brought an extremely complex subject to life with dramatic visuals and made the information understandable to a variety of disciplines in the audience.

SEEING IS BELIEVING

Engaging the sense of sight as well as the sense of hearing doubles the effectiveness of communication. If I read you a description of a panther, it might take you several seconds to identify what I was talking about. But if I showed you a picture of a panther, your brain would be able to interpret the visual message instantaneously.

We receive 1 percent of our information by taste, 1.5 percent by touch, 3.5 percent by smell, 11 percent by hearing, and 83 percent by sight. Eight hours after a presentation, audiences will have forgotten most of what they heard if the information received was not accompanied by visuals.

Visuals are particularly helpful when addressing nontechnical audiences about technical subjects, when speaking to audiences with multiple levels of knowledge, low levels of literacy, or international audiences. A presenter can use visuals to increase involvement in the sagging middle of a presentation, when audience attention is usually the lowest.

FOLLOWING A PROCESS

Many of my clients tell me that they don't know where to start when creating visuals to accompany their presentation. It is useful to follow a *defined process*.

IDENTIFYING IDEAS TO TRANSLATE INTO VISUAL IMAGES

You have already determined your objective for the presentation. You know what response you want from the audience. You have prepared and organized your material. Now select items and concepts that would benefit from being made more understandable or memorable through visuals. *20 percent of what you say will make 80 percent of the impact.* This critical 20 percent should be in visual form; you can add more, but do so sparingly. Your visuals are not meant to contain every word or idea of your speech.

Your visuals might include:

- agenda

- nature and scope of the material

- comparisons, contrasts, or abstract ideas

- critical statistics and financial figures

- large data sets

- personnel, equipment, and budgetary requirements

- methodology

- a timeline

- sequential information

- detailed design or operation

- geographical data

- results of your research

- conclusions

- recommendations and next steps.

QUESTIONS TO ASK BEFORE YOU CHOOSE THE MEDIUM

Once you have selected the information that's best suited to visual representation, you'll need to decide on the appropriate medium (platform or output) to display your visuals. There are several different types of presentation media you can use, ranging from low tech to high tech. The selection of the best medium will depend on many criteria, including the following:

1. What are your predesignated specifications or constraints?

Constraints arise from a variety of factors. The audience may have predictable preferences. For example, government RFPs will often specify exact requirements. The environment and facilities may restrict your choices. Your level of personal comfort and expertise may also set limits.

2. What resources are available in your company?

Become acquainted with the equipment and staff that your company provides. Seek out others who have done similar projects and ask for assistance and cost-saving tips. You may have access to a graphics department designer who will turn your chicken scratches into exquisite visuals and deliver them ahead of schedule. More likely, your graphic design department may reside in your PC, and you are fully responsible for the content, design, and delivery.

3. What is your budget?

Some speakers visualize a Disney production without considering time constraints and, more importantly, budget constraints. It is essential to understand the importance of this specific presentation to your company. This importance is usually defined by the time and money allotted to the project.

Defining a financial budget will thrust you into reality. Since visuals reflect both your personal and your company's image, create the best *quality* visuals you can afford.

Set up a basic budget and track actual costs. As soon as you go beyond your own capabilities for an electronic presentation, expect to spend more. Be alert to deceptive savings. A cheaper, but less experienced supplier may cause delays or not produce quality work.

Examine low, medium, and high cost options and spend your money to get the best return for your investment. It might be wise to use a simple laptop presentation with generic templates and an LCD projector, and use your money to give your audience professionally designed four-color, bound notebooks to take with them.

An engineer told me it is a mistake to "spend enough to lose" when responding to a proposal. His firm either does it well enough to achieve their objective or not at all. They intend to respond less often, spend less, and win more. For example, a public works director invited them to make a proposal for a project the following week. "We don't expect full-blown graphics," the director said, "just something 'quick and dirty' will suffice. This is an opportunity for us to get to know you." Preliminary questioning revealed that another engineering firm was virtually assured of the job. Upon further investigation they also discovered the city would put a $2 million job out to bid the next month. Their objective and budget quickly changed. Instead of showing up with a few plain graphics, they spent a thousand dollars on charts and a computer-generated presentation. The presentation introduced team members that would be well suited for the project a month out. Although they lost this first project by one point, the city felt the graphics were so well done, they asked to buy them. The engineering firm gave them to the city free of charge and are now better positioned to compete for the upcoming project.

4. What is your timeline?

What can you *realistically* accomplish in the *time* available when you factor in all your other professional and personal responsibilities? What lead times are required?

How long will it take to create visuals for the different types of mediums? Are they elaborate or simple? Will you need to work with internal or external resources? What are the milestones? Conferences may expect abstracts and handouts four to six months in advance. Deliverables and reliance on other people create scheduling complexity.

Develop a project schedule with the deadline at least two to three days before the actual presentation to give you time to rehearse, edit, rehearse again, and assimilate visuals into your delivery. Work backwards to realistically fit in target dates. Even if you have a short lead time, having a priority checklist will ensure that you have covered all the bases. If you have multiple elements for a complex multimedia presentation, it will demand a schedule!

Allow for crisis. A client found it took her longer to get clearances and permissions for the visuals for a satellite videoconference than it did to produce the two-hour show. After months of letters, phone calls, and faxes, she was still on the phone fifteen minutes prior to broadcast getting authorization for the opening sixty-second video roll-in.

5. What are the physical realities of the site?

You may have to prepare your visuals without knowing anything about the room in which you will be speaking. For example, some professional conferences assign rooms at the last minute. This requires flexible planning. Whenever possible, you should demand the following information:

- Size of audience

- Room size; it will dictate size of visual images

- Distribution of audience in room: visuals must be legible for entire audience

- Equipment available including size of projection screens, monitors or projection equipment, audio equipment, and computers

- Personnel required to operate equipment

- Adequacy of power sources

- Lighting

- Portability requirements.

Are you going to be traveling and working in a variety of settings that will require you to carry self-contained units? Will you design different kinds of visuals for different kinds of settings and equipment? Or will you create one set of visuals for the lowest technological level? You may find that even though you brought all your equipment for an international presentation, there isn't enough electrical power.

6. What are the preferences and prejudices of the audience?

Determine your audience's

- Expectations

- Sense of the appropriate

- Dominant learning style(s)

- Facility with the English language.

Colorful and well-designed visuals can create a strong emotional response—either favorable or unfavorable—to your objective. For this reason, your audience analysis should be reviewed again, as it will help you determine specific needs, tastes, and hot buttons. It will also help you decide what quality of visuals will engage your audience and bring about the response you want.

Base all your decisions on how you want your listener to react. You may opt for viewgraphs if you want to create an intimate "down to business" feeling at a monthly project update for senior management. You can have normal lighting, maintain better rapport and eye contact, avoid equipment hassles, and be more in control. However, it may be *de rigueur* in your industry to use the latest high tech, even for in-house presentations. Analysis can reveal that your audience of conservative investors would object to a complicated electronic presentation as a wasteful use of their money. Perhaps a short, simple computer-generated slide show would get your message across and also show fiscal responsibility.

Your choice of medium will also be influenced by secondary audiences that will reference your presentation, including Web access or audio/video tapes. If you are being audiotaped and are using several props and detailed photos, either describe them carefully or include a handout with the tape.

7. How proficient are you in operating the medium of your choice?

This is a critical question. Be brutally honest in your answer. If you are not comfortable with the equipment, you will need to build enough extra time in the schedule to both create the presentation and *master the equipment.*

Depending upon the answers to the above questions, you should be able to narrow down the choice and select a visual platform.

CHOOSING THE APPROPRIATE MEDIUM

As you consider your options, do not eliminate a medium simply because it is inexpensive or low tech. Your message may come across more effectively and you may be perceived as more competent with a simpler medium.

Flip Charts and Whiteboards

Before you totally disregard this section, consider a recent assignment. I was asked to work on the preparation of a bid for a multimillion dollar government contract by URS Greiner and subcontractor CH2MHill. I helped coach a two-member engineering team presenting a two-hour oral proposal. The ground rules stated that you could generate your graphics on a computer, but the use of computers, TV screens, LCD screens, or any computer-run devices were not allowed. No overhead transparencies or handouts were allowed. Flip charts, foam boards, diagrams, and photos were permitted.

The presentation was to be videotaped by a stationary camera. This influenced the size of the visuals. The presenters were used to presenting with slides, viewgraphs, or computer-generated slides and dimmed lights. It would be a challenge to gain and keep the attention of the evaluating panel for two hours. The material was extremely complex. The graphic designer made two oversize colorful charts detailing the cleanup site, and several flip charts with photos and text. Actually, the large, colorful charts turned out to be a memorable way for the panel to study the before and after renderings of how the firm intended to carry out the proposed contract. It took many rehearsals to choreograph the use of the two flip charts and smoothly put up and take down the large charts. But the presenters were able to come across as in control, maintain eye contact, and keep the emphasis on rapport with the audience.

A few hints may be helpful:

- Use flip chart pads with faint blue grid lines to help keep your letters and diagrams aligned.

- Check that pages in the entire flip chart have not been used before.

- Start with a fresh sheet, especially if a previous speaker has used the flip chart.

- Bring your *own* colored marking pens; use dark colors to print, underline, and make arrows.

Choosing the Appropriate Visual Medium

Flip Charts—up to 30 people

Advantages
- Inexpensive
- Interactive, casual
- Normal lighting
- Spontaneous
- Can customize
- No equipment problems
- Short lead time
- Can post for reinforcement
- Control

Disadvantages
- Limited visibility
- Requires good handwriting
- Static images
- Low-tech image

Viewgraphs—up to 100 people

Advantages
- Flexibility
- Inexpensive
- Interactive
- Good eye contact
- Short lead time
- Colorful
- Presenter has control
- Portable
- Few equipment problems

Disadvantages
- Low to normal lighting
- Glaring light
- Difficult to keep clean
- Noisy
- Low-tech image

Slides—up to 1000 people

Advantages
- Professional
- Portable
- Colorful
- Sharp detail
- Long
- Easy to store

Disadvantages
- No flexibility
- Can be expensive
- Limited interaction
- Low lighting
- Limited eye contact
- Can detract from speaker

Video—depends on screen size

Advantages
- Attention getting
- Normal lighting
- Easily duplicated
- Motion, animation
- Compresses information
- Shows real people, places, objects
- Involves audience emotionally
- Special effects

Disadvantages
- No flexibility
- Expensive
- Limited interaction
- Limited eye contact
- Long lead time
- May overshadow speaker
- Difficult to transport equipment

Electronic—depends on screen size

Advantages
- Flexibility
- Highly interactive
- Normal lighting
- Massage information
- Import data
- Latest information
- Animation
- Printouts available
- Archivable
- Portable
- High-tech image
- Special effects

Disadvantages
- Initial costs expensive
- Needs projection system
- Learning curve for production and delivery
- Rehearsal mandatory
- Equipment failure
- Presenter can be seen as technician
- Requires adequate power/ communication sources

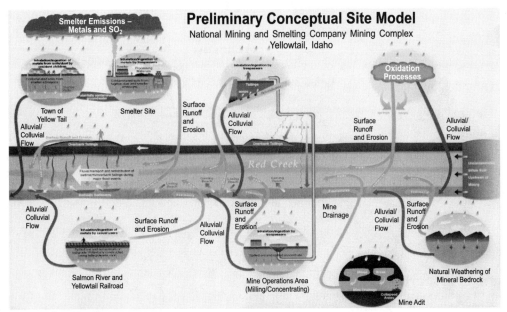

Courtesy of CH2MHill

- Review the correct spelling of terms that may be used.

- Write important facts, figures, or major headings on the chart before your presentation.

- Write in plain, large letters that are *legible* to everyone in the room.

- Avoid changing the wording or intent when writing down ideas from the audience.

- Practice flipping the sheets unobtrusively.

- Tape sheets around the room for reinforcement of ideas and for later reference.

- Two charts are helpful if you wish to make comparisons.

Electronic whiteboards print instant black and white or color copies of anything you write or draw for distribution to attendees. Some have PC interfaces so you can add graphics, or fax or e-mail copies to other individuals.

Whatever equipment you use, *do not neglect your audience.* Write or draw briefly, and then turn and reestablish eye contact.

Viewgraphs

Viewgraphs are sometimes referred to as viewfoils, overheads, foils, or transparencies and are shown on overhead projectors. In theory, these low-tech visuals are slowly being relegated to the era of the slide rule, but actually are found

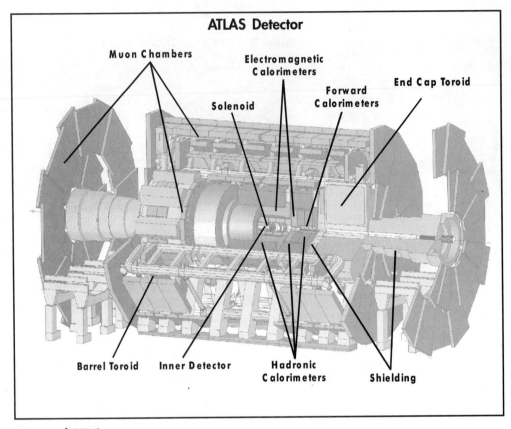

ATLAS Detector

Muon Chambers

Electromagnetic Calorimeters

End Cap Toroid

Solenoid

Forward Calorimeters

Barrel Toroid   Inner Detector   Hadronic Calorimeters   Shielding

Courtesy of CERN[2]

alive and functioning in many companies, especially in science and engineering. A conservative audience will react favorably to them. They are cost effective, easy to design and change, and require simple, portable equipment. This is especially critical for presentations in third-world countries. More importantly, a presenter can feel in control of the visuals and can focus on the audience.

- Strip off the protective sheet of paper *before* you get in front of your audience.

- Use plastic covers to keep viewgraphs clean.

- Zoom in on small details of your viewgraphs.

- Increase the timeliness of your presentation by clipping a relevant headline or article from a current industry magazine and creating a viewgraph.

Slides

35 mm slides have sharp images and text with intense color. Bio-tech, medical, and engineering presenters will find the ultra-fine resolution of slides desirable when details and accuracy are essential. Dramatic effects can be created by

using multiple projectors and combining them with sound effects or music. Slide projectors are simple to operate and reliable. Your computer graphics program can be linked by modem to a slide service and you can have low-cost slides in a short turnaround time.

Organize the sequence and technicality of slides to suit the audience's mental framework, not your own. For example, you may need to add a few extra steps even though you are familiar with a process. Many audiences will expect a printout of the slides included in their handouts.

- *Check that your slides are inserted properly in the slide tray and secure the cover.* Run a colored marker along the top edge of the slides when they are loaded to indicate they are right-side-up. Then if you don't see a color mark, you know the slide could be upside down.

- *Conduct a trial run with your remote control and preview every slide.* If you are one of several speakers and need to set up your equipment in front of the audience, check the focus and center the image on the screen. Place masking tape on the floor to indicate where to correctly position the projector table.

- *Check the level of light necessary for good visibility.* Rear projection (light projected from behind the screen) will allow you to have low-level light in the room. Avoid being a disembodied voice by keeping adequate light on yourself when the rest of the room is darkened.

- *Group your visuals to avoid switching the lights on and off frequently.* For example, begin your presentation by speaking directly to the audience, show several slides, then turn up the lights for further discussion. Repeat the process two or three times; then finish with the lights up for the question and answer portion. Appoint someone to manage the light switches.

Video

Audiences, especially younger audiences, are conditioned to pay attention to video. Video is an excellent medium for involving your audience emotionally. Video is especially useful for on-location filming of equipment, processes, or people.

- Mention in your introduction *when* the video will be shown if monitors or a large screen are present.

- Divert attention from a small monitor by facing it away from the audience until you are ready to use it.

- Use video clips to add variety and change pace in a presentation.

- Brief the audience on key points or images they are to look for during the tape.

- Emphasize important data with freeze frames.

- If you have more than twenty people in your audience, you will need a video projection system and a large screen.

ELECTRONIC PRESENTATIONS

You may chose to create an electronic presentation or perhaps your company policy or audience expectations dictate the use of this medium. You may be a first-time presenter or you may be a seasoned veteran. Technology can trip up smart people who want to give smart presentations. However, if you do master the technology, there are many benefits.

Benefits

It is extremely difficult to get and keep an audience's attention when you are presenting factual, dry information. Electronic presentations that are highly stimulating and interactive can help hold the audience's attention. For example, you can conduct "what if" experiments by manipulating images and photos on the screen as the audience raises questions, offers opinions, and makes suggestions.

Electronic visuals can also help audiences with varying levels of knowledge visualize complex technical ideas that are difficult to describe in words or static visuals. You can use electronic visuals to compress information that would take a long time to verbalize into visual forms that you can display in a few seconds. Sequential information, detailed design, and operational procedures will all benefit from animation, clip art, and special effects. Electronic presentations can clarify information for international audiences and also increase the retention of your material.

It can be very difficult to sway an audience who resolutely cling to their beliefs, no matter how silver-tongued you are. Bob Betz, Vice-president of Stimson Lane, who craft Ste. Michelle wines, was faced with vendors and customers who insisted that French and California wines were better than Washington wines. But Bob is a myth-breaker. His basic premise was that Washington State had similar soil, rainfall, and climatic conditions to France and California, and therefore could produce grapes and world-class wines. Betz, who has a background in agri-science, went to a multimedia producer and together they designed maps of the three regions detailing rainfall, terrain, and growing conditions. His animated graphics sweep the viewer over mountains and contours, descend through simulated fog and mist to the vineyards, and illustrate the similarities of Washington with California and France. It is difficult for his audiences to admit that they have been wrong, but Betz presents irrefutable proof to his audiences. Statistics are convincing, but the visual enhancements of the statistics convince everyone in his audiences.

If you learn something new about your subject, situation, or audience at the last minute, you have the ability to modify your visuals or download the most up-to-date information from the Internet. In addition, material saved digitally is easily available for posting on a Web site or publishing in print.

> *"Visuals used to be a way of freezing reality;*
> *now they are a way of thawing imagination."*
> —Unknown

Technophobia

Many people feel so incompetent with high-tech equipment that they fear losing control of the process. With good reason! Equipment failure can divert even the experienced presenter's attention away from the message and the audience. If you make infrequent presentations, you will not only have to reestablish a comfort level, but will probably also have to adjust to new technology. Since we have all seen speakers embarrassed by equipment failure, it is easy to visualize a similar catastrophe happening to us. We fear looking like a fool or being reduced to the role of an incompetent technician.

The PR director of a bio-tech company told me they simply avoid this possibility:

> Despite all the technology, we, as a matter of policy, just use a whiteboard or, at most, a 35 mm slide projector. Beyond that, the technology risks detract from the presentation and its overall impact. Science and engineering professionals are seduced by technology. The learning curve and the rehearsal time needed for an accomplished speaker to create a smooth presentation are far

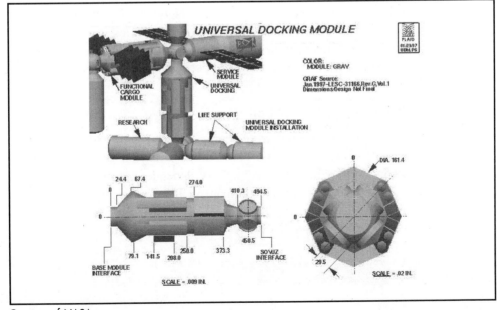

Courtesy of NASA

too long. An inexperienced speaker attempting to use sophisticated technology is a disaster waiting to happen. The audience becomes impatient with the glitches. I've witnessed so many industry presentations in which twenty different companies give pitches. Invariably, someone stumbles. The price can be very high.

Strategies for Managing Technophobia

These fears are real and require a strategy, including more thorough preparation at all levels. Gaining skill in using electronic equipment and preparing effective presentations takes time and effort.

Sometimes you may be expected to go high tech without adequate preparation. Many scientists report that the main criteria for promotion in their company is based on periodic project updates on their work, and that they are asked to provide high-tech visuals, even though the company refuses to provide training in the sophisticated software needed. Since they want to project an image of expertise, they must dedicate a disproportionate amount of energy to preparing visuals. As a result, they sacrifice time and effort that should be put into content. I advise people in this situation that they need to obtain outside basic training on the software. They should give priority to building content and establishing a progressive storyline incorporating their interpretations and recommendations, and then spend time on constructing a *few good quality* visuals.

- *Keep in mind that the more technologically complicated you make your presentation, the more vulnerable you are to disaster.*

  In addition to thorough preparation, you need to be aware of and control all the surrounding elements that can affect your presentation. What are the power sources, how fast can you download files from the Internet, do you require an ISDN (Integrated Services Digital Network) line? Does your equipment have adequate memory? Do not take facilities, room size, and audience size for granted. Often you will be working with people who do not understand that this information is important. *Keep asking* until you get the information you need.

- *Use a minimum of technology to accomplish your objectives.*

  Audiences may equate the failure of your equipment with the failure of your product or service. A backup plan will restore your credibility as well as your product's. If you're too technology dependent, you risk losing creativity, humor, perspective, and humanity. The primary purpose of all visual aids is to *clarify the meanings of your ideas*. If they fail to enhance your ideas, don't allow them to clutter up your presentation.

- *Avoid sabotaging yourself.*

  Check the basics: loose cables, on-off switches, an unfamiliar laptop, unexpected glare, or a repetitive noise. It's OK to start your presentation with humor as long as the joke isn't your equipment. It is *not* endearing for a speaker to confess to the audience, "I don't know a thing about running this stuff."

- *Demonstrate that you are in control of your technology.*

  For example, make your point *before* clicking to the slide that illustrates it, while the previous slide is still up. If you show the slide and then make your point, it will give the impression that the slides are running the show. You might say, "Let me draw your attention to..." or "I'd like to review my recommendations for further tests...."

- *Practice with new equipment to acquire skills before you are under pressure and facing the deadline of a project update or proposal.*

  Reviewing a videotape of your rehearsal will quickly reveal how smoothly you handle remote controls, mouse, or laser pointer and how much time you spend looking at your laptop screen. Keep the emphasis on your audience.

Production

A computer-generated slide show has unique production and delivery elements that are not the same as a slide show presented through a carousel projector. The first difference is that you are now much more involved in the production of the visuals. Put on your "producer" hat as you need to see and choreograph your presentation as an overall production. You now may have:

- More people involved

- Bigger budget

- Complicated equipment

- Expanded timeline that calls for project management.

Depending upon the size of your laptop, you may be able to display your visuals to as many as five or six people at one time. For up to ten people, you could use a desktop or TV monitor. Beyond that you will need an LCD (liquid crystal display) or CRT (cathode ray tube) projector to project images on a larger screen. Sharpness of the images will depend upon the power of your equipment, the background, and the lighting. Robert Lindstrom, author of *Businessweek Guide to Multimedia Presentations*, suggests, "Measure the distance from your farthest seat and divide it by 8, which gives you the desirable height of your image." For example, if the back row is sixteen feet away, then your image must be two feet high to be visible.

Portability—Pack Your Own Chute

Traveling compounds problems. Even if your company provides a technician, you still need to be familiar with the basics, and be able to ask critical questions of hotel and conference staff.

Equipment

The basics are similar to any type of visual. Interfacing between your equipment and the equipment available on site becomes critical. This means either bringing everything with you, or arriving with enough time to check everything out so that your computer doesn't crash when opened. It is simpler for small audiences to use just a laptop and a power source, but a bigger audience will require a data projection system to transmit the images to a large screen.

Personnel

Establish contact with a technologically sophisticated individual in charge. Your presentation may depend on his or her expertise. Unless you are traveling with a crew, you need to have many conversations, faxes, or e-mail with this person and know that he or she will be on-site during your presentation. Neglect this and you may be shot down by a 70's sound system, unavailable or incompatible power sources, or a guy in the control room moonlighting from Joe and Moe's TV repair.

Here are some suggestions for when you are "on the road"

- A self-contained projector unit is a good travel alternative to an LCD/ overhead projector combination and is also brighter.

- Take more than a read-only viewer version so you can update or customize your presentation.

- Check to see if your company's insurance policy or your personal policy cover damage and theft of your electronic equipment. Whereas not many people would run off with your overhead projector, laptops are not only portable for you but for a thief. Last year the theft of notebook computers exceeded $1 billion and the lost data was worth even more. Take precautions—especially in airports—to keep your visuals and laptop from vanishing.

- Label everything as fragile and use the protective carrying cases recommended by the manufacturer. The vibrations of constant travel can shake things loose.

- Pack powerstrip, extension cord, tape to hold cables, pointer, extra lightbulb for projector, and an extra laptop battery.

- Check out your presentation on your notebook and the external display upon arrival.

- Bring a copy of your presentation on disk so if equipment fails, you can give your presentation on rental equipment. You may need a copy of your graphics viewer program in case it's not loaded on a rental PC. In case of emergency, *have the number of a local, reliable rental service that will deliver to your presentation site.*

- Learn troubleshooting basics to get equipment and software up and running.

- Bring a low-tech version of your visuals.

On-site activity should be minimal. Concentrate on your content, interact with the audience, relax, and enjoy the presentation.

Recycling

You don't always have to start from scratch to produce an electronic presentation. A multimedia production firm or in-house department can reformat your existing slides or video into a laptop show. As you analyze your audience, you may find that altering the background, scanning in recent photos, and adding a brief animated sequence may be the only customizing required. Perhaps the content can use some updating, but essentially remains the same.

Look for ways to recycle your visuals after you've spent so much valuable time and money on your presentation. Pool your resources by collaborating with other presenters within or outside your department. It may be useful to combine your computer-generated slide show with their video, and add narration and sound to produce a multimedia program for a second project. You could burn your visuals to CD and mail them to absentees or contribute them to your annual report. Make your visuals available on your intranet or company Web site.

USING STORYBOARDS

Storyboarding is an excellent tool for you to get your ideas on paper, translate them into images, and decide on the best placement of visuals in your presentation. Many people find it useful to start a storyboard as they begin their audience analysis. They incorporate the audience's needs and hot buttons and add to this visualized outline as they determine their main and supporting points. It is also very helpful as they work out an organizational pattern. It provides a way to build on a progression of ideas from the beginning to the middle to the end. You will be able to see your presentation in its entirety.

Your storyboard can be drawn in many different formats. The object is to see everything at once and be able to change its position. You can use ordinary 8.5- by 11-inch sheets of paper and adjust and edit them on a table or a wall. Or you can condense your storyboard onto a single sheet of paper and use two- by three-inch grids or smaller. You can even make one out of post-it notes and use different colors for main points, supporting points, and visuals.

Capsulize your ideas into short statements. Develop the content of your storyboard by using key words and sketches. Jot down some titles. An engineer told me her company uses storyboards to clarify objectives and to ensure continuity and seamless transitions in team presentations. "We just stick everything up on the wall. That way every member of the team can check out content and visuals of other team members early in the game."

Storyboards will help you sense the pace of your information. A structure will begin to form as you adjust the visual outline of your presentation. You may discover that you need another visual to clarify a certain concept or that you can eliminate a visual and still convey the necessary information. A storyboard will help you balance stimulating portions with necessary but less exciting data. If you are producing an electronic presentation, a storyboard is especially helpful to coordinate elements of visual effects, animation, sound, and video.

I introduced the storyboard concept to one of my clients who was designing visuals for an hour presentation. His storyboard revealed fifty-four computer-generated slides and every one of them was text. Too many and too boring. He omitted seven slides. I pointed out that he should not have more than three consecutive visuals consisting of text, so he replaced 20 of the text slides with

Storyboard

images or images and text. Once the slides were altered, he was able to review and edit the slides with the slide show tool on his graphics program. The storyboard still needed more variety, but his lead time was too short to develop anything more complex. I suggested that he show thirty slides, then shut off the computer, turn up the lights, and talk directly to his audience. When he came to his conclusions and future recommendations, he could return to his visuals to highlight the points he was making. Later he told me that he felt gaining rapport with the group in the middle of his presentation helped final acceptance of his proposal.

The brain loses its ability to function if overwhelmed visually as well as verbally. A good rule of thumb is to have one visual for every two minutes of your presentation. Keep in mind, however, that even though you could use fifteen visuals in a thirty-minute speech, you may only *need* five. An animated sequence or a two-minute video may effectively illustrate your entire message. *Spend a minimum of 40 percent of your time in direct contact with the audience without visuals.* Storyboards are *your* visual aids in preparation. They are particularly useful for showing whether you are allocating a proper portion of the time to essential points of your message.

VISUAL DESIGN

Now that you have your preliminary sketches for visuals, a plan for their use during your presentation, and the medium, it is time to decide on a functional design for each visual. Here are some design strategies that will aid your audience's speed of comprehension, recall, and understanding of the quantitative information on your visual.

- Use graphics to present dominant conclusions or features.

- Choose simple typefaces. Typefaces carry a message of style. Serifs (finishing strokes) are more classic, add more authority and formality, and can be easier for continuous reading, whereas sans serifs present a more clean, modern, and functional image.

- Make text *legible* and large enough to be *visible* to your entire audience.

- Limit text to five words per line, four to six lines per visual.

- If it is necessary to give an overall view of a complicated system or process, show enlarged details on a second or third visual.

- Include both *text* and *images*. Words alone take time for the audience to read, process, and interpret. Strong images make data stand out. Images should be easily recognizable, provocative, unique, funny, or unusual to trigger associations and aid retention.

- Include a persuasive point or benefit in your title. Check out newspaper and magazine headlines to see how they capture the essence of an article.

- Position information on the top three-fourths of the visual for back row visibility.

- Emphasize points with underlining, boldface, color, italic, or reverse type.

- Label the plot storyline clearly.

- Use *upper-* and *lowercase*-type instead of all capitals, which are harder to read.

- Check carefully for grammatical or numerical errors and misspelled words.

A good rule is: As Simple As Possible (ASAP), but not simplistic. *Sometimes more detail can clarify.* If the information on your slide can be dispensed within twenty-five to thirty seconds, perhaps there is not enough substance, and you should question the value of including it. Merge the information with another slide. Edward Tufte, author of the *Quantitative Display of Visual Information,* advises to:

> Find design strategies that reveal detail and complexity—rather than to fault the data for an excess of complication. Or worse, to fault the viewers for a lack of understanding.

Tufte suggests techniques of layering and separation, visually stratifying various aspects of the data.

- *Don't gild the lily.* The editing floor is cluttered with some of the best dramatic cinematic moments of world-famous actors that never made it to the movie screen. Why? Because an astute editor cut scenes that were too long, too boring, or too much. The result—crisp, progressive stories that hold your attention. You don't need as much self control as a movie editor, because anything you trim from your presentation *can* be used another time.

- *Personalize your visuals* by including names, facts, or photos depicting members of the audience. Include a logo or symbol representing the group you are addressing.

- *Choose clip art that reflects the world and its diversity of cultures.* Decide if cartoons will lend a trivial quality to your message, or should you opt for realism? Perhaps adding a photo from a digital camera, scanner, or photo CD would be more appropriate for your audience/topic.

- *Change, add, prioritize, and eliminate unnecessary details* that appeal to you but may have little meaning for your listeners. Maybe some details would be more useful as a handout.

- *Avoid overusing elaborate wipes and transitions* between computer-generated slides or during video unless you are producing MTV. They can easily distract attention from your message. Professional film editors will make hard cuts 90 percent of the time or dissolves to black; flashy transitions rarely add to meaning.

- *Be prepared to spontaneously navigate* to any portion of your material in response to audience feedback.

I like to ask my clients, *"So what?"* as they display each visual. *"Will this image clarify or prove a point?"* I also ask, "What if you don't include it?"

Icons

Visuals can condense large amounts of data into concise images. One of my favorite features in the newspaper is *Earthweek: a diary of the planet.* Symbols placed on a map of the world summarize information from the U.S. Climate Analysis Center, U.S. Earthquake Information Center, and the World Meteorological Organization. The map records all the unusual phenomena occurring during the week. Earthquakes, tidal waves, tropical storms, snakes, or unusual insect populations, global-warming alerts, monsoons, or El Niño are all depicted with icons. We can read scattered items in the newspaper, hear about them on the radio, or even see isolated stories on television, but the visual of *Earthweek* has a much stronger impact. When you can see earth movements that took place during the week in New Zealand, Japan, Pakistan, Venezuela, and Mexico, you suddenly become aware of all the interconnections of the ecosystem.

Choices of Graphic Format

Choosing the best format that quickly clarifies your information will take some thought. You may want to compare a new process with a current process, establish a pattern, or show relationships between two operations. Will this

Pie Chart, Bar Chart, Line Graph, Flow Chart, Table

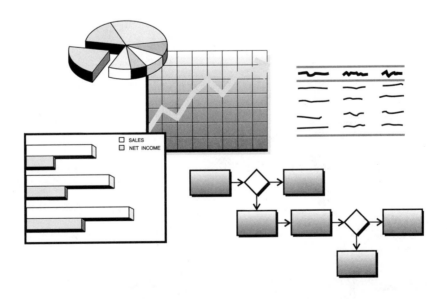

information be clarified best if described in text, a diagram, a table of statistics, a flowchart, or a bar chart?

Do you have any photographs available to scan into your program or can you incorporate a digital photo? Would it be helpful to include an animated sequence in the middle to add variety and interest? Perhaps you can choose one or two illustrations and then include supplementary material in your handouts.

Translate your information into formats such as:

- *Tables*—show a large amount of data, and relationships between items.

- *Pie charts*—circles divided into sections to show component parts such as proportions or percentages.

- *Bar charts*—highlight similarities, contrasts, ranks, proportions, and frequencies. They have a single common variable. They can have either vertical or horizontal bars.

- *Line graphs*—depict changes in one or more variables, trends, increases/ decreases, or concentrations over a period of time.

- *Flow charts and process charts*—show various stages of a process or illustrate the line of command in an organization, or relationships among parts of a structure.

- *Maps*—show spatial relationships and can be used to illustrate the geographical distribution of resources, products, sales figures, populations, warehouses, and so on.

- *Photographs*—show actual details of objects, places, or persons.

- *Cartoons*—can simplify or exaggerate data in a humorous way.

- *Drawings*—illustrate ideas with images.

Cartoon, Map, Drawing, Photo

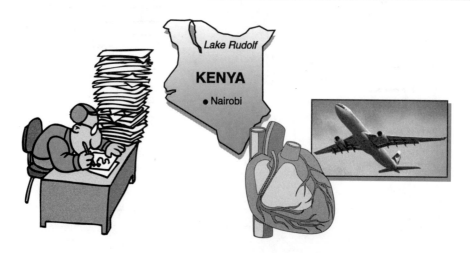

Color

Color influences us on both conscious and subconscious levels. Its emotional impact can affect the perception of ideas in a presentation. Colors provide consistency to unify your message, or contrast to introduce new subjects or emphasize differences. The fact that you have a choice of 16,000 color hues does not mean you have to use them all. The trend is away from the garish primary color schemes of earlier presentation software programs. Muted colors are not only kinder to the eyes, but a designer can incorporate more information with only slight shade variances.

Your choice of medium should influence your choice of colors.

- If you have low resolution, such as with an overhead projector, use a clear background and dark lettering.

- On an LCD panel, use high contrast colors for background and text.

- Slides have higher resolution and, therefore, you can use a wider range of shades.

Conduct a color test on your output platform to see if colors have the same sharp contrast as on your computer monitor.

Use color to emphasize. Use accent colors as a highlighter to call attention to key ideas, financial results, or critical data.

Use color to organize material. The same color template will provide consistency throughout a presentation, but you may wish to use different colors for emphasis. For example, you could use one color scheme for the problem, another for the solutions, and another for future actions.

Use color to link information or to trace processes.

Remember that color affects emotions. For example:

- Blue is a cool, soothing color that suggests trust and a conservative approach.

- Red is a stimulating color that suggests danger, passion, and action.

- Yellow is a warm color that suggests enthusiasm and liveliness.

Audio

Sound is usually an afterthought, yet creative use of sound, music, and expressive speaking can heighten attention, enhance emotion, and add emphasis to your message. An unobtrusive musical track underneath spoken words can immediately set a mood.

Handouts

Handouts, which are commonly duplicates of visuals, are expected in some organizations. It may be necessary to include complicated charts, schematics, and statistics in handouts for you to convince key decision makers. Developing your handouts will help you organize your speech and improve the flow of material during your delivery. Well-prepared handouts provide structure and an organizational pattern for both the audience and you to follow. They can explain in detail, provide supplementary material, references, case studies, a glossary of terms, and be a record of the presentation. Including a brief biographical sketch and picture of the presenter on the final page will help establish credentials.

It is always difficult to decide whether to distribute handouts before or after the presentation. There is no easy answer. If your purpose is to communicate information, it may be more useful to hand them out first. If your purpose is to lead your audience to a decision, it may be best to distribute the handouts later so that the listeners' attention is not diverted. You may choose to pass out the most critical information at the same time you present it. One engineer told me that he prepared exact duplicates of his slides as handouts and passed them out prior to his presentation. The key decision maker at his meeting flipped through the handouts and, tossing them aside, said, "Your numbers don't add up. It's not worth doing!" The engineer's entire presentation had been reviewed and dismissed, and he never had a chance to present his material adequately.

James Baltusnik, former programs officer for the U.S. Federal Park System, says, "Since my material in my presentation is complex and the group knows that they must try to understand, they are ill at ease. I tell them I will give them a handout at the end with a summary. I find there is too much confusion if I give handouts during a meeting, but if they know they're getting a handout at the end, they aren't so obsessed with taking notes." If you are energized, have a strong voice, and have reasonable pacing, it is possible to keep control and focus the audience's attention on you, a visual, or a prop. It is when the sense of progression is lost that the audience wants to read ahead.

If your handouts make sense on their own, you may want to make them accessible from your Web site. Supplementary or updated information can also be provided in this way.

The appearance of your handouts should be in keeping with the high quality of your visuals and the rest of your presentation.

- Include relevant clip art, diagrams, and sidebars to break up complex information.

- Color code pages or add tabs to help your audience quickly find pages you reference. Trainer Bob Pike prints his handouts on yellow paper and then tells the class they can look it up in the "yellow pages."

- Leave ample white space for notes or have blank pages for "keepers" or "action steps."

- Bind handouts to indicate the contents have lasting value, as well as to add to your professional image.

- Carry a master copy that can be duplicated if necessary.

PROPS

Displaying a real object can explain an abstract concept and lend credibility and drama to your message. Alan Levy is CEO of Heartstream, a division of Hewlett-Packard that manufactures a portable heart defibrillator. He immediately captures the attention of his audience by stating that, "350,000 people will die of cardiac arrest this year and it will most likely be the cause of *your* death."

Then he explains ventricular fibrillation by using his hands to create a visual analogy. "The heart is a pump," he begins. He places his fingertips together in a prayer-like fashion. "All of my fingers represent the muscles of the heart. They are all receiving the same signals." Levy flexes his fingertips together in a rhythmic pattern to demonstrate the organized contractions of the heart. "However, if the signals become erratic and chaotic," he continues, "the muscles contract in a disorganized manner and the heart stops pumping blood." He illustrates by making random spasmodic movements with his fingers. "The only cure or therapy for this abnormality is to apply an electrical charge to the chest. The shock actually stops the chaotic rhythms. Then the heart restarts in a normal manner." Levy abruptly stops the movement of his fingers and then reverts to the original regular movements. "It's like rebooting a computer."

Levy talks about the need for a portable, easy-to-operate device that is safe, requires no maintenance and minimum training. He keeps his product hidden until he asks, "So what is the solution?" Then he produces the four-pound heart defibrillator. The audience is amazed because it is a fraction of the size they expect. He asks for an audience member who is nervous about using a defibrillator to come to the front of the room and be part of the demonstration. The simple-to-operate unit is a very persuasive prop.

- Keep your props out of sight until you need them to illustrate a point.

- Make sure they are clearly visible to everyone in the audience.

- Pass an object around if you have a small audience so they can experience properties such as weight, smoothness, or hardness.

- Put props away when they have served their purpose.

- Invite your audience to view detailed props or models after the presentation.

INTEGRATING VISUALS WITH DELIVERY STYLE

Now that your visuals are completed, you need to incorporate them smoothly into your delivery. Your choice of visuals will affect your speaking style and comfort level. If you want to come across as confident and competent, do your facial expressions belay your distraction and inadequacy when you have problems with electronic equipment? I was coaching a senior executive on an important presentation to be delivered the following week. Although he usually exhibited complete confidence and control when he presented with viewgraphs, he repeatedly had trouble with a laptop computer and the time it took to download or switch between programs. When the laptop monitor went blank and he had to turn to the screen to see what, if anything, was being projected, he yanked the electrical cord from the wall and declared, "Forget hi-tech! I'm using viewgraphs!" I agreed, and told him it would be best to rehearse with his laptop when there wasn't the pressure of a critical presentation.

The information and mood of the visuals should match your words, facial expressions, body language, and tone of voice. Your audience will receive the same information through many channels. Don't make an important point while you are doing something physical, such as changing viewgraphs or accessing files on a computer, because the audience will give more attention to watching you or the computer screen than to listening to you. Be careful of nervously playing with a pointer, mouse, or remote control.

Face the audience. Place your laptop or controls in front of you. A computer stand will position the laptop at waist level so you can stand up straight. The normal tendency when showing visuals is for the speaker to turn away from the audience to focus on the screen or the equipment. If you are using an overhead projector, step away from the projector frequently and reestablish rapport with the audience. Turn it off when you are not using it.

It is always preferable to have the *screen on your left* as you face the audience. People read from left to right and their eyes will return to you on their left. Give the audience a moment to absorb the information on a visual, then repeat the key ideas and talk briefly about the subject. Don't look at the screen again unless you need to point to something. *Keep your eyes and attention focused on the audience, not on the visual aid!*

Audience attention is highest at the beginning and end of your presentation. It lags considerably in the middle. If you don't have many visuals, this may be the time to show them and revive interest. If you have a long, involved visual program, this is the point where you should change the pace by turning up the lights and find out if you are "on the same page" as your audience. Based on this feedback, you may end up eliminating parts of your program. There is absolutely no reason for showing inapplicable slides just because you created them!

Ask for a white matte screen with a keystone eliminator (a steel bar that pulls out of the extension tube and flips forward), which will allow you to hang the screen at a slant and get a good square image.

The screen should be placed in the corner at a slight angle to the audience so the presenter will not be walking in front of the screen and obstructing the view. However, many technical facilities place the screen squarely in the center and the presenter needs to avoid walking through the light beam of the projector. A pointer can be useful, but don't allow it to become a distracting toy. Laser pointers require a steady hand. If the screen is large, you may be some distance away. A strong voice will pull the audience's attention back to you.

In the chapter on *Designing Your Finish First,* I suggested memorizing the first and last few sentences of your presentation so you can make strong, stable eye contact with the audience. This will do more to build and maintain your credibility than any other physical action. Establish this rapport with your audience *before* you turn on the overhead projector or computer program. Allow time in the ending to summarize and reestablish the emotional connection with the audience. Prepare some ad-libs to help you keep your composure in case of crisis with the equipment.

Technical Rehearsal—Be Prepared

Your decision to use visuals, regardless of the medium, will necessitate *extra rehearsal* time. Since visuals are often completed at the last minute, many presenters have no opportunity to rehearse with them. They forfeit the value of rehearsal and the opportunity to review and revise. If the visuals aren't finished, use mock-ups to modify and adjust your pacing according to time spent on showing each visual, waiting for build sequences, the realistic demands of turning flip chart pages, passing out handouts or props, downloading materials, establishing Internet connections, and other activities. Rehearsal will allow you to properly focus projectors and check legibility of each visual, look for spelling errors, see that the viewgraphs are clean, cue up videotapes, check slides to see if they are inserted properly, and rehearse manipulation of computer files and software.

If this is a presentation for clients or outsiders, rehearse in front of diverse groups in the organization to get feedback about content. Perhaps they can suggest the use of a graphic, image, or chart that would highlight specialized information that you aren't aware of or may have overlooked. If you choose a complex electronic presentation, you *must* plan for extra rehearsal time in your preparation schedule. If this extra practice is not realistic, simplify or forgo the visuals.

Become familiar with your sequence of visuals, using a computer printout or storyboard, so that you can concentrate on interacting with the audience. Visuals can be a crutch, and many people hide behind them. You are not incidental to the presentation—*you are your own best visual.*

VISUALS CAN ENHANCE—THEY CAN'T SUBSTITUTE FOR YOUR MESSAGE

Several years ago, I was teaching government personnel in Anchorage, Alaska, how to prepare presentations. A CIA administrator walked to the front of the room with a stack of viewgraphs. He began by looking at the empty screen. "This picture would have clarified my first point," he said, "but I can't show it to you because it's classified." He held up a second viewgraph without revealing its content and said, "You would be able to see how this graph sharply illustrates the difference, but you'll have to use your imagination." He then picked up the next viewgraph, saying, "This picture shows vivid details of the project, but it's classified too." The audience waited in anticipation and uncertainty as he proceeded to refer to imaginary images on the screen throughout the rest of his presentation. The substance of his talk had vanished with the visuals.

Your visuals may not be classified, but if your equipment failed, would your audience still be able to get the necessary information from your words? Would you be able to achieve your objective?

EMERGING TECHNOLOGY FOR ELECTRONIC PRESENTATIONS

It is difficult to even speculate about future advances in presentation technology. The reader will need to continually adapt the basics in this chapter to the changing technology. New and emerging presentation tools are making it easier to create exciting visuals and making the exchange of information more interactive. Meeting rooms are equipped with computer stations at each seat, allowing members to electronically converse with the presenter, who can access their responses through a control panel (usually a touchscreen) built into the lectern. Auditoriums and lecture halls with remote keypads in the arms of the chairs allow members to interact with the presenter, who uses a lectern with a control panel and a computer. The computer also provides the presenter with audience responses to questions, both for the whole group and individually.

As interactivity increases, the presenter is becoming more of a participant in computer-enhanced communication. Communications consultant Bernie DeKoven foresees: "Everybody—fellow participants, the person playing emcee, the person playing technographer (controlling computer and data)— will control the pace and focus of the presentation. The 'live' environment of the shared computer screen allows speakers to make very powerful, responsive presentations."

Many presentations are currently delivered via desktop. Your graphic presentation may be downloaded from the Internet while you walk observers at remote sites through the material via the phone. Whereas in a face-to-face meeting the presenter is the most important visual, you are now no longer able to give the verbal and physical cues that could help clarify your message, and it is harder to accurately gauge audience response. The structure takes on a greater significance when a presenter can be eclipsed. An expressive voice becomes an influential, persuasive tool.

During a class at NASA Lewis Research Center, excited latecomers explained they were tardy because the visuals of a presentation had been so fascinating. A contractor had ended his presentation with a shimmering hologram of a proposed facility. No one mentioned the presenter.

The bar continues to be raised for production standards as software programs make sophisticated sound, 3-D animation, and photo-realistic graphics available to novice users.

Michael Stephan, graphic staff consultant for CH2MHill, used a very persuasive electronic presentation to help a prospective client visualize alternative designs:

> To display what a project will look like when complete, we often create several alternatives to spark discussion on which alternative is best. The first image is an original photograph of an existing two-lane road. The next image shows a photo simulation of the two-lane road, expanded to include a left-turn access lane and intersection enhancements. In the third image the road is expanded to four lanes, including a left-turn access lane. Alternatives display the various size relationships of the designs and the impact to the surrounding environment.

Road Photographs

Courtesy of Michael Stephan

Many presentations are being videotaped or burned into CD-ROMs. Immersive photography and virtual reality enables viewers to not only see pictures, but literally submerge themselves in them. These enticing, comprehensive, and interactive images can be very persuasive, but require a presenter with superior communication skills.

The Web will eventually be used routinely for presentations. You will be digitized into permanence and instantaneously available on demand to worldwide audiences. If you want to gain significant visibility, master this technology. This desktop learning medium can provide synchronized high-quality digital video, audio, text, and presentation slides.

• An expert can give one live presentation and have it archived for future reference.

• Users have access to presentations at their convenience. They can navigate through one or many speeches or search by key words to select relevant issues for their perusal.

- The variety of formats can provide a deaf person with visual text or a visually impaired person with audio of the presentation.

- Multiple versions of the transcript in various languages are invaluable when dealing with global audiences.

- Management can track usage of the Web site, conduct testing, or get detailed feedback from users.

To prepare for Web presentations, follow the suggestions for on-camera appearances in the chapter on videoconferencing. You and your visuals will be subject to opinionated scrutiny. This exposure sets higher standards for presenters; it would be wise to thoroughly prepare and rehearse.

THINK IN IMAGES

Today, we are dealing with more information than we can comfortably absorb. Visuals are a way to condense and pace that information. Effective visuals help you present complex information in a logical, planned sequence to influence and facilitate problem solving and decision making. The visually sophisticated audiences of today are accustomed to high-quality visuals.

Visuals are attention-getters or attention-losers. Use visuals only when they add value to the message. Emphasize human contact and interaction, rather than presentation technology. No matter what kind of high-tech tools you choose to use, you should not forget the original purpose of your presentation: *to deliver a message to your audience in the clearest, most easily understood manner.*

KEY IDEAS

- You are your own best visual!

- Choose and design visuals that will influence the audience emotionally, as well as intellectually.

- Use a variety of image graphics, as well as text.

- Rehearse with all equipment and have a Plan B.

- Support and enhance your message with visuals, but don't let them dominate the presentation.

Notes

1. All visuals in this chapter were originally produced in full color.
2. The Large Hadron Collider (LHC) being built at CERN in Geneva, Switzerland, will collide protons into protons at a center-of-mass energy of about 14 TeV. When completed in the year 2005, it will be the most powerful particle accelerator in the world. It is hoped that it will unlock many of the remaining secrets of particle physics. The LHC accelerator facility will contain the ATLAS detector, which is about the size of a five-story building.

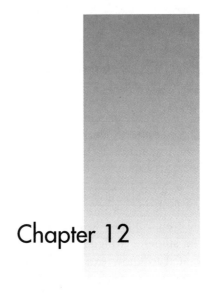

Chapter 12

ADDING VARIETY

"The spectator forgives everything except dreariness."
—Voltaire

OVERVIEW

The only thing worse than sitting through a boring presentation is giving one! This chapter encourages you to lift your presentation out of the mundane by adding variety to your material, voice, style, and the way you interact with your listeners. You must capture and keep the attention of your audience to communicate your message and accomplish your objective. Retain the listener's attention by using contrasts, changing stimuli, and alternating the mood within your presentation.

Bill Nye, star of the television program, "Bill Nye, The Science Guy," is an example of a model communicator who consciously orchestrates his presentations. He considers teaching to be a performance and uses stories, analogies, props, visual aids, and demonstrations to communicate principles of science to his audiences:

> First of all, I try to think of innovative analogies for the ideas I am teaching. For instance, I wanted to answer a popular question: "If you are in an elevator that starts falling, can you save yourself by jumping up at the last moment?" I realized this was an opportunity to explain potential and kinetic energy. I designed a Barbie- and Ken-size wooden elevator about 9 inches tall in an 8-foot-high elevator shaft. My passenger was an egg. I reeled the

egg up to the top of the elevator shaft and explained that this was potential energy. This potential energy was converted to kinetic or moving energy when I cut the rope and let the egg go. A spring mechanism caused the egg to "jump" a second before impact. The smashed egg clearly answered the initial question. If you fall 500 feet in an elevator and are able to jump up 12 inches, you still fall 499 feet!

A local physics professor told me he tries to stimulate his students' thinking and look for ways to move his classes beyond the theoretical. For example, he illustrated magnetic levitation by using a new high-temperature superconductor. The students watched him demonstrate this extraordinary phenomenon and wanted to know exactly how it worked. His advice to speakers is to "play your audience and fan a small interest into a fire."

Just as you would become bored listening to a repetitious melody for an hour, an audience's attention will wander if your material, pacing, and interaction do not vary. In today's electronic environment, our focus is on speed and hyperefficiency. Audiences have become used to the computer's nanosecond time frame and your information must now compete with, and triumph over, other stimuli. MTV and programs such as *ER* and *NYPD Blue* use edits that are less than 1/24th of a second and barely have time to register on the conscious level. Nancy Cook from Boeing Center for Leadership and Learning stated that "Employees expect 'Edutainment.' By that I mean training must be delivered in an entertaining manner that is close to professional production standards." Today's audiences expect that you will not only inform, but stimulate and entertain them.

Usually an audience will listen attentively for the first ten to twelve minutes of a speech. Thirty minutes into an hour-long speech, the audience's attention is at its lowest ebb. After fifty minutes, it is difficult for an impatient audience to concentrate on any material, no matter how fascinating the subject is for the presenter. Keep in mind that, according to one study, an executive's attention span maxes out at six minutes.

An audience listening to scientific and technical information doesn't possess a miraculous trait that improves their concentration. In fact, when you present complex information, unfamiliar concepts, extensive background material, statistics, and a myriad of dry details, it takes more energy for your audience to focus, search for associations, and store the information in long-term memory. It is important that you give the audience periods within your speech to relax and assimilate the information. Paul Cook, trainer for TCI cable company, says he expects his trainees to take "mental vacations" during two- and three-day seminars on highly technical material. He begins the session by describing the destination for his personal mental escapes and then asks participants to share their "getaways." "It's a fun way for me to get to know the class, and they learn about each other," he says. "It immediately starts the interactive process, which I work hard to encourage throughout the training."

Keep in mind that time seems to pass more slowly for the audience than it does for the presenter. A model communicator takes responsibility for keeping the audience alert and consciously adds variety to enliven the presentation. A preacher had a reputation throughout the countryside of being a dynamic speaker. An aspiring young clergyman sought him out and asked for his advice. "I understand that you keep everyone's attention during your sermons," he said. "How do you do it?" "Well," replied the preacher, "it's quite simple. I have a young man with a long, pointed stick who stands on the platform with me. Whenever anyone in the congregation starts to fall asleep, he pokes *me.*"

Now that you have prepared and organized your material, review it to see where you can add some variety. Look at your material and your choice of words. Do you have an abundance of words at your disposal to precisely and vividly describe your thoughts? How can you add variety to your delivery, especially your voice? Can you increase the interaction with your audience?

THE ULTIMATE WORD PROCESSOR—YOU

> *"Language is a cracked kettle on which we tap out crude rhythms for bears to dance to, while we long to make music that will melt the stars."*
>
> —Gustave Flaubert

A golfer selects the appropriate putter, a fisherman the right lure, a carpenter the proper tool. If you select vivid, descriptive words and use them precisely, you will have more control over the visual images that others call up in their minds. An excellent style cannot be achieved with a small vocabulary. Your words should be concrete, unambiguous, economical, and specific. The use of "buzz words" can indicate you are familiar with current industry trends. However, be aware that overuse of a buzz word can turn it into a cliché, such as "re-engineering."

Model communicators are known for their originality and individuality of expression, accurate vocabulary, and command of grammar. Do your words obscure or enhance your meaning? Do your words have spirit and energy? Will they call up vivid associations and fire off the neurons in the minds of your listeners?

The average person's speaking vocabulary is very small. Researchers recorded telephone conversations of approximately 80,000 words and found that speakers had a working vocabulary of only 2,040 words. Elevate your speaking vocabulary to the level of your reading vocabulary. Keep a copy of *Rodale's Synonym Finder* or *Roget's Thesaurus* on your desk at all times. Many software programs include a thesaurus to help you find that perfect word. Increasing your vocabulary is tantamount to increasing your thinking capacity.

Have you considered the connotations as well as the denotations of your words? The denotation is the explicit meaning, and connotations are what the word suggests. The name of the Patriot antimissile used by the coalition forces during the Gulf War connoted freedom, the flag, and loyalty. The mere thought of the word "SCUD" for the missiles from the Iraqi forces suggested the "bad guys." Use words that evoke rich and favorable imagery.

Usually, technical speeches are impersonal, and presenters use third-person pronouns. Because the research is seen as separate and distinct from the researcher, the speaker slips into the passive voice: "The bridge was determined to be unstable..." or "The cultures were subjected to heat." The impersonal voice is also overused: "It should be noted that..." or "The research demonstrates...." The sentence structure, as you can see in these examples, becomes cautious, circumspect, and loses its forcefulness. The verb can be your strongest element in a statement. Active verbs require concrete, interesting nouns. Make them specific, bright, lively, and vigorous. Streamline your prose and *choose active rather than passive verbs.* "The engineers noted that the bridge..." or "We heated the cultures..." Identify who is doing what with first- and second-person pronouns: "To analyze the data, I fed the samples through..." or "You should note...."

If you speak in abstract generalities, you are giving your audiences more leeway to think about other things. If your audiences have to stop and figure out what you're saying, they lose contact with your message. Help them avoid blind alleys and running into walls by using a colorful vocabulary that illuminates your ideas. If your words are vague, they may be open to several interpretations and may weaken your message. Be specific: take responsibility for what you say.

Words should not be used to gratuitously demonstrate your knowledge or to impress people. William Butler Yeats cautioned, "Think like a wise man but communicate in the language of the people." Sometimes the professional buzz words and jargon are equated with the elitist aspects and complexity of a profession, and members of that profession are reluctant to relinquish this mystique. Empower your audiences by giving them new word choices. They won't be alarmed or confused if you introduce new words, define them carefully, and emphasize the words as you use them.

Replace unnecessary technical words with a word familiar to your specific audience. For example, a physicist might talk about coherent superposition, whereas a chemist refers to synergy, but a general audience would understand that the whole is greater than the sum of the parts.

Choose emotionally powerful words. As author and science editor Edward Tenner points out in his book, *Tech Speak:*

> Most dictators are direct. They use simple honest words like home, children, blood, soil, work, fate, strength, youth, and, of course, victory.

> But scientists have retained from the seventeenth century a profound suspicion of eloquent argument. Where lawyers see no injustice in having a case decided for the party with the more articulate champion, scientists believe that evidence, not

persuasion, must prevail. Of course, this leads some of them to a
kind of reverse rhetoric that uses needless complexity to imply
an absence of verbal tricks. Here again, language reform has led
to what is denounced as language abuse.[1]

Publishing consultant J. Wendell Forbes reminded an audience that in the
age of scientific approaches, "The gentle art of communication is of paramount
import. Nothing cuts through the information clutter better than a well-turned
phrase." Your choice of words should bridge the familiar to the unfamiliar. Your
words should unify rather than separate people.

Variety in Delivery and Interaction With Listeners

How can you add variety and contrast to your delivery? If you remain behind
the lectern, your lack of activity will lull the audience to sleep. Very few immo-
bile people can keep the attention of their listeners. James Burke, host of the
British Broadcasting Company programs *Connections* and *The Day the Earth
Changed,* is one exception. His eye contact, humor, drama, and rapid-fire
eloquence make each person feel that Burke is talking individually to him or
her, even when there are 2,500 people present! A speaker needs a commanding
voice if he is riveted to one spot. When there is no discrimination with the voice,
the audience is forced to make their own interpretation of what is important and
what needs to be remembered. Vary your voice by speaking loudly and softly,
changing pitch, and using pauses.

Moving into the audience can get their attention. Wait until you sense you
have achieved rapport and it is okay to invade the audience's space. Cross the
invisible barrier two or three times during the presentation and the question and
answer period. If you are on a platform and need the height to be seen, you can
still go down the steps into the audience briefly. Practice these moves so they go
smoothly.

The speaker sends information to the audience, gets back information, and
adapts or modifies her material and delivery. *If you aren't receiving information
from your listeners, communication, in essence, has broken down and won't
work to its maximum efficiency.* One engineer confided to me, "The audience
distracts me and, if I get too involved with them, I lose my place." I encouraged
him to pay attention to reactions from his audience, for they will forgive lapses
of memory before they forgive being bored or ignored.

Nowadays, *an audience expects interactivity.* Ask your listeners rhetorical
questions, or ask questions that require them to raise their hands. In some
situations, you can have your audience write down answers to questions and
then share them with a partner or within a small group. Find out the names of
your audience, put them on tent cards, or have people wear name tags.
Announce early in your presentation that you will be looking for feedback later
on, then walk into the audience and call people by name. "Dave, have you ever
had the problem of...?" or "Sarah, how do you feel about...?" Asking your

listeners about their feelings is nonthreatening because there is no right or wrong answer. The activity will keep them alert. Be cautious about asking a very personal question and expecting an individual to share the answer with the rest of the group. For example, I ask my audience to think back to a time when they had a fearful speech experience or felt humiliated, but I ask them to change the ending in their mind. No one has to discuss it aloud unless someone volunteers to do so.

Your audience understands and learns not only by seeing and hearing, but by feeling, smelling, tasting, and touching. Use tactile images and words such as "wet," "ice-cold," "slippery," or "rough" to describe physical attributes. A pilot described what g-force felt like by asking his audience to imagine they had to make every movement with sixty-pound weights on each arm. Some presenters play upbeat music as their audience files in for an association meeting or use background music to enhance the mood of slides or a video. All of these add interest to presentations.

Jack Vallentyne, a Canadian ecologist who performs for children under the name of Johnnie Biosphere, demonstrates to his audience how everyone is a part of the ecosystem. He asks his listeners to hold their breath. He then tells them that the gas molecules now in their lungs have been in the lungs of every-one else in the room. He further emphasizes that we are linked with every other human being from Jesus Christ to Michael Jackson by shared molecules in the air, water, and soil. Sometimes the audience tries to stop breathing because they are shocked and even disgusted by the idea that they are sharing the air with others. This graphic demonstration involves the children mentally, physically, and emotionally. They will always remember we are all part of the same ecosystem.

To Read or Not to Read...

It is a rare speaker who can read from a manuscript and still be successful at injecting variety into his voice, style, and audience interaction. However, some technical audiences prefer that you stick to a manuscript, especially when there are complex formulas and critical information. Before reading from a manu-script, you should plan several rehearsals to acquire familiarity with the material so that you can concentrate on the audience. Double- or triple-spacing your manuscript will make the words easier to read. Underline key words and draw wavy lines over important phrases where you should change your inflection, add emphasis, or pause. As you read aloud, you will find sentences that are unwieldy, too long, or complicated. Break them into manageable segments and create a more conversational style. Think the thoughts as if you were saying them for the first time. Speech writer Jerry Tarver says, "When spoken words lack a proper beat, listeners smell the odor of ink."

STYLE—MORE THAN THE SUM OF ITS PARTS

In *Wind in the Willows*, Toad's egotistical style of speaking forced his friends into blunt honesty:

> "Now, look here, Toad," said the Rat. "It's about this Banquet, and very sorry I am to have to speak to you like this. But we want you to understand clearly, once and for all, that there are going to be no speeches and no songs. Try and grasp the fact that on this occasion we're not arguing with you; we're just telling you..."
>
> "Mayn't I sing them just one *little* song?" Toad pleaded piteously.
>
> "No, not *one* little song," replied the Rat firmly... "It's no good, Toady; you know well that your songs are all conceit and boasting and vanity; and your speeches are all self-praise and—and—well, and gross exaggeration and—and—"
>
> "And gas," put in the Badger in his common way.
>
> "It's for your own good, Toady," went on the Rat...
>
> "It was, to be sure (Toad said), but a small thing that I asked, merely leave to blossom and expand for yet one more evening, to let myself go and hear the tumultuous applause that always seems to me—somehow—to bring out my best qualities. However, you are right, I know, and I am wrong... But, O dear, O dear, this is a hard world!"
>
> And pressing his handkerchief to his face, he left the room with faltering footsteps.[2]

Toad had to face the fact that style wouldn't lift his speeches out of mediocrity if there was not substance. But style and substance are not mutually exclusive. The mistaken perception in many scientific and technical arenas is that to have substance, you have to give up style, and vice versa. Layne A. Longfellow, a dynamic model communicator, emphatically states that the "more profound and intellectual your message, the more you need to emphasize style."

Everyone has a personal style of delivery. Your style is a composite of the words you select, your gestures and movements, appearance, voice, and the way you make your ideas available to others. A columnist once wrote of President Kennedy: "It is no use trying to say what I mean by the Kennedy style. Style is not something one can define exactly or prescribe for another. It has something to do with taste, something to do with restraint and control, and something to do, finally, with grace and gallantry." Movie actress Raquel Welch declares, "Style is being yourself—on purpose."

Stanford University professor and author John W. Gardner comments, "The image makers encourage the individual to fashion yourself into a smooth coin, negotiable in any market." Rather than fading into the woodwork in your organization, you can—and should—insist on expressing your unique personality in your presentations. What sets you apart? Think of small creative ways in which you can add interest to your style, and remember that rarely will anyone protest if you present a dry subject in an entertaining and informative way.

Be innovative. Increase the frequency of your eye contact with people in the audience, and add more intensity to your voice. Smile. Be open and vulnerable. Use humor and crispness. The tempo of music changes—so should the pacing of your presentation.

Your style should change to suit the situation, the material, and the audience. A lighthearted after-dinner speech for your local engineering association will require a more casual style than a formal budget request. Your language, delivery, pacing, and appearance may all change.

Your ideas may be beneficial and your facts accurate, but if you lose the attention of your audience, you will not get the response you want. Go back over your material and examine where you can vary material and the pacing of your delivery. Think of ways to present your information vividly, forcefully, and economically in your own inimitable style.

KEY IDEAS

- Keep the audience's attention by introducing change throughout your presentation.

- Help your audience make associations by using descriptive, concrete words.

- Develop and trust your own unique style.

- Do not fear that a strong personal style is incompatible with substance.

- Adapt your style to your audience and the situation.

Notes

1. Edward Tenner, *Tech Speak* (New York: Crown Publishers, Inc., 1986).
2. Reprinted with the permission of Atheneum Books for Young Readers, an imprint of Simon & Schuster Children's Publishing Division from *The Wind in the Willows* by Kenneth Grahame. Copyright 1908, 1933, 1953 Charles Scribner's Sons; copyright renewed (c) 1961 Ernest H. Shepard.

Chapter 13

REHEARSING

"You did say you wanted to be out of the mob, didn't you?"
"Yes, but how did you...?"
"Like everything else, Fletcher. Practice."
—Richard Bach
Jonathan Livingston Seagull

OVERVIEW

Rehearsal, both physical and mental, frees you from internal critiques and allows you to concentrate on your words and feedback from the audience. This chapter describes how practice will improve your delivery, make you feel secure, and help you adapt to unexpected circumstances. It suggests a checklist and schedule for rehearsals and criteria to evaluate rehearsals and performance.

Rehearsal is an important final step in the process of mastering your material and developing a sense of timing. If you feel comfortable, you won't panic should something go awry. Most people know they should prepare and rehearse presentations, but many speakers tend to emphasize the accuracy of their material and postpone attention to delivery until the last minute. One engineer said, "It is always on-the-fly around here. Often I am told to give a briefing for the following morning. I don't have the opportunity to do anything more than a few simple graphics on my computer. I'm lucky if I have time to review the material in my head while I'm on the way to work."

Information changes so rapidly that speakers often feel uncomfortable trying to explain it to others because they haven't had the time to become familiar with the latest data themselves. It requires extra effort on the part of the technical professional to find the time to prepare thoroughly and to do a run-through. The speaker's reward for preparation and rehearsal is feeling more comfortable during the presentation and eliciting genuine applause at the conclusion.

PHYSICAL REHEARSAL

Why rehearse out loud? Because it is an efficient and powerful way to edit and revise. The German word for rehearsal is *die Probe,* and indeed it provides an opportunity for the speaker to probe and examine, and to check out material and delivery.

Set up your rehearsal area as closely as possible to the physical proportions of the actual speech site. Whether you are confined to a small boardroom in close proximity to your audience or speaking to 500 people seated in an auditorium, you will get more value out of rehearsing under simulated conditions.

It is important to walk through the speech from beginning to end. I was videotaping one client and asked her to present her material *exactly* as she would give it the next day. She began, "And here I do this, and then I'll show a photo, and then I'll give the next example." I stopped her. "Don't talk about what you'll do, do it. If you get under pressure, you will revert back to how you rehearsed it." "You're right," she agreed. "In my last presentation, I actually said, 'and then I'll walk over to the flip chart.'"

If at all possible, visit the actual speech site. This will give you "*home court advantage.*" Every sports team wants to play on their home field or court because they know what it looks like, how it smells, the feel of their feet on the floor or grass, and the sound of the fans. They don't have to process extra information so they can concentrate on playing the game. No actor would dream of getting on the stage without weeks of rehearsal. Actors always want to rehearse on site with their props to assimilate all the physical details and perfect their timing.

Sit on the chairs, touch the tables, and count the number of steps from your seat to the front of the room. Notice how the paint is worn thin on the lectern where previous speakers have gripped it with white knuckles. Look everywhere. Listen to the sounds from next door, from outside, or the drone of air conditioning. If you intend to use a pointer, rehearse with it; and practice putting it away! Walk around the room making large gestures. Keep your body in comfortable alignment. Your gestures and body language will come easily and spontaneously because of this practice.

You may encounter problems that you can't resolve. Feedback from other speakers who have appeared before the same audience can be a valuable resource. Call and ask them about the general attitude, knowledge level, types of questions, and any problems that they experienced at the site of the speech.

Graphics are usually finished at the last minute. You will integrate them more successfully into your presentation if you practice with preliminary sketches and check the time necessary for you to review, and the audience to read and interpret, your slides or viewgraphs. Practice the sequence of visuals while directing your attention toward the audience. It is mandatory to actually hook up your laptop and practice advancing your computer-generated slide show.

If you will use a lectern, rehearse with one, or stack some books on a table and put your notes on top. It makes a difference in your eye contact with the audience. If you have to be letter perfect with your script, check out portable speech-prompting devices that work much like TV TelePrompTers. They are simple to operate, but require practice to develop a comfortable speed and smooth delivery.

If you are the third or fourth presenter, practice walking to the front with your materials and changing or adjusting the audiovisual equipment. Even if your meeting is a regular in-house presentation, rehearse getting up from your chair and going to the front of the room, smoothly setting down your laptop or viewgraphs, reaching for the proper switch to turn on the projector or computer, and distributing handouts.

Rehearse that humorous story or anecdote in the actual room, or one of the same size. It may go over well in an intimate office, but fall flat in a large ballroom. One of my clients had to speak at Radio City Music Hall after the Rockettes had performed. I reserved Seattle's Fifth Avenue Theater, which seats 2,500, so that my client could get acclimated to a large space. Later, he reported this hour-long rehearsal was a major factor in his ability to "own his space" in Radio City Music Hall.

Friends and colleagues may point out your weaknesses, but may not be objective or helpful in suggesting what would improve your delivery. I find my clients can watch a videotape and realize where they should make some basic changes. Start videotaping your rehearsals as you get close to the presentation date. It will not do you any good to watch yourself fumbling through the presentation. In fact, I like to fast-forward through the rough spots and focus on the strengths of the presenter. Psychologist Peter W. Dorick is a proponent of this self-modeling theory that people learn from their own successes—not from negative feedback. Dorick carefully edits videotapes and only lets his subjects view themselves being successful in carrying out tasks. He believes this method has helped individuals gain confidence and master skills faster.

Notes

When someone says that he will speak "off the cuff," we know that he will be speaking impromptu, with no preparation. Originally, "off the cuff" meant that the speaker wrote notes on his shirt cuff that could be easily read without giving the appearance of having notes. Experienced speakers have always known that a good ad-lib takes hours of preparation.

Memorize the first few lines of your speech for an impressive opening. Practice directing your attention and energy toward the audience. Know the route you are traveling onstage and where you will be during each part of your presentation. Memorize the final few lines of your speech and the second ending after the question and answer period.

Read through your speech again and concentrate not only on the main points, but on the details supporting each main point. You can use 5x8 cards to help organize your information, but be sure to number them. Give your speech using whatever words happen to come to mind. Get through the whole speech, even if you have to gloss over certain parts. Note how long it takes. When are you at the halfway point? When are you five minutes from the end? Take into account that during your actual speech you may talk faster if you are nervous.

The night before your speech, list the main headings of your presentation and read them aloud several times. A model communicator told me he tape-records his speech and listens to the tape on his portable tape recorder during the flight to the city where he'll be speaking. Another client told me she has a precise routine she follows on the day of a presentation. She always gets up at 5 a.m., dresses, and does a complete run-through. She is motivated to practice, she explained, by the memory of how confident she felt the last time she gave a rehearsed presentation. Try different behaviors until you find a routine that works for you. One engineer rehearses several times and then refuses to even think about his presentation the day before. He feels this "breather" allows him to sound fresh and spontaneous.

If you are chosen to present an important proposal, you will usually be expected to rehearse in front of other team members who will suggest changes to facts, statistics, and visuals. This feedback is necessary and can be extremely helpful halfway through the process. But if team members continue to tear apart your presentation up until the last minute, it can be very detrimental. One person should assume the role of the director and all suggestions should be filtered through this person. The presenter will become confused if three or four individuals are giving conflicting directions. It is important that your material is locked in at least two to three days in advance so that you will have a chance to assimilate the information and "make it your own." I have actually seen a presenter look at a graphic in bewilderment during the final presentation and later tell me that someone had added new bullet points he had never seen before.

The *chemistry* between a selection committee and the presenters can often be the deciding factor when awarding a contract. They need to feel that you will be comfortable to work with. But if someone has just demanded ten changes the night before, and you haven't had an opportunity to rehearse them thoroughly, your delivery will reflect your unfamiliarity and uncertainty. If you are not in control of the presentation, how can the committee expect you will be in control of the work in the contract? Insist all new input stops on a certain date and that further communication with you must be in writing.

At a certain point, rehearsals must cease. Time should be allotted to prepare your mind and your body. If you are severely fatigued and your energy is depleted, it will negatively affect the recitation of the most accurate and compelling statistics. The content must have substance, but your *rapport with the audience, commitment, and control are paramount.*

MENTAL REHEARSAL

Visualization is simply mental practice, the process of forming a mental image of your presentation. The mind does not distinguish between imagining a situation in detail and actually doing it. Many athletes who compete at the international level recognize the value of mental practice.

Doug Reynolds, an Olympic pistol marksman, says he imagines himself going through each step of his three-minute preparation period, in which he methodically lays out everything he will need during his performance. The process allows him to enter each match with an automatic rhythm and a better chance to adapt to the things that will inevitably go wrong.

Some people find it difficult to visualize. If this seems foreign to you, give it a try, because it does work successfully for many people. Take some quiet time. Close your eyes, sit back, and imagine the entire situation. Use all your senses. What will you hear, see, smell, touch? See yourself *successfully* giving the presentation. Visualize the audience responding favorably to your material. The mental image of success must be implanted firmly in your mind before it can occur in reality.

In *The Cerebral Symphony,* neurophysiologist William Calvin observes:

> As you become highly practiced in a skill, it no longer qualifies as novel. It does move from conscious control to less-than-conscious. In Zen archery, the object is to become so practiced that no conscious will is required to release the arrow at the right moment. You can simply watch the arrow being released, as if someone else were doing it.[1]

This quiet time will also give you a chance to pull all your scattered energies together. If you are experiencing stress in your professional or personal life, you may be emotionally fatigued. This will be apparent in your voice and body language. Mental rehearsal can help restore your mental stability and confidence so that you are not vulnerable to the land mines of a high-pressure situation.

DRESS REHEARSAL

Actors are used to switching costumes as they switch roles. They make a conscious effort to rehearse the body language appropriate to the costume. For example, if an actress has a role in a period play, a good director will insist on her wearing a long skirt from the first day of rehearsal. Her stride will become more graceful, her posture more erect, and the actress will enter a room differently than if she were wearing blue jeans.

You too can benefit from a few dress rehearsals. This is particularly true if you are used to casual attire, as many lab scientists and high-tech computer personnel are. Spend some time in your suit or best casual dress, and discover the mannerisms that come from being constrained by a belt, tie, a collared shirt, or straight skirt and uncomfortable dress shoes. Become the executive staring back at you from the mirror. Assume an authoritative posture. Talk at yourself to find the voice that goes with your power outfit. You *can* come across as genuine! We all have different facets within us, but you may not have been in touch with the particular characteristics that are required for this specific communication situation.

REHEARSAL SCHEDULE

When basics become habit, a speaker can concentrate on the audience, adapting to the feedback he receives, or to any unexpected circumstance. If you are worried about your content or appearing nervous, you could find yourself overwhelmed if a senior executive arrives unannounced, or if extra chairs and handouts are suddenly needed for additional attendees. You can win points with the audience by smoothly handling interruptions or adverse conditions. Experience, of course, will make you feel more at ease. But rehearsal is the next best thing. Following is a suggested rehearsal schedule. The evaluation checklist at the end of this chapter can also be used to critique a videotaped rehearsal.

Preparation

1. Create a mind map or brainstorm for the essential points to be covered.

2. Research important facts, statistics, and proof.

3. Make a rough outline.

4. Write a first draft—write the finish first.

5. Translate key concepts (20 percent) into visuals. Do rough sketches of more visuals; create a storyboard.

1st Rehearsal

1. Start seeing speech as a whole and decide on an organizational pattern.

2. Walk and talk through your presentation from beginning to end.

3. Include sketches of visuals.

4. Double-check accuracy of all statistics, facts, and technical data.

2ⁿᵈ Rehearsal

1. Give presentation. Check time, props, and visuals.

2. How can you involve audience mentally, physically, emotionally?

3. Add variety through examples, analogies, and humor.

4. Finalize site and equipment arrangements.

5. Choose attire that is comfortable and professional. Wear during practice.

3ʳᵈ Rehearsal

1. Videotape or tape-record presentation. Use minimum of notes. Check time. Rehearse with others if a team presentation.

2. Get feedback and modify as needed.

3. Prepare handouts—get necessary clearances.

4. Check audiovisual equipment; practice with microphone.

5. Memorize opening and ending of speech.

6. Anticipate questions and jot down answers.

Presentation

1. Do a last minute check on appropriate clothing and grooming.

2. Warm up your voice; hum. Do physical exercises to get rid of tension.

3. Eat and drink lightly.

4. Have some quiet time by yourself to recharge your energy.

5. Concentrate on your audience.

6. *Enjoy yourself* and give a dynamite presentation!

Try to cover as many of the above points as possible, even when you don't have much lead time. Expect that each situation will be different. Be flexible so that you can adapt the subject and your style to the particular audience and situation.

Recently, my son and I dialed my sister in California and immediately launched into an enthusiastic duet of "Happy Birthday." When we finished, an unfamiliar voice responded, "I think you've got the wrong number." I was profusely apologizing when the listener interrupted, "No problem. You really needed the practice, anyway." Eventually everyone does rehearse. *Make sure your rehearsal is not before an audience!*

EVALUATION CHECKLIST (VIDEO OBSERVATION SHEET)

Mark each statement: X for "excellent," S for "satisfactory," ✔ for "needs improvement."

A.   Organization and Development of Content

Opening statement gained immediate attention?   _____
Purpose of presentation made clear?   _____
Previewed contents of speech?   _____
Main ideas stated clearly and logically?   _____
Organizational pattern easy to follow?   _____
Main points explained or proved by supporting points?   _____
Variety of supporting points (testimony, statistics, etc.)?   _____
Conclusion adequately summed up main points, purpose?   _____

B.   Delivery

Presenter "owned the space" and was in control?   _____
Held rapport with audience throughout speech?   _____
Eye contact to everyone in audience?   _____
Strong posture and meaningful gestures?   _____

C.   Visuals

Visuals clear and visible to entire audience?   _____
Creative and emphasized main points?   _____
Presenter handled unobtrusively and focused on audience?   _____

D.   Voice

Volume
Rate (pacing)
Pitch   _____
Quality   _____
Energetic and included everyone in dialogue?   _____

E.   Comments

KEY IDEAS

• Start rehearsing entire speech early.

• Prepare to speak "extemporaneously."

• Memorize the opening and closing lines of your speech.

• Rehearse at the actual site of your speech.

• Visualize a successful presentation.

Notes

1. William H. Calvin, *Cerebral Symphony* (New York: Bantam Books, 1990).

Part III

MAKING A COMPELLING DELIVERY

Chapter 14

COMMUNICATING WITH
EFFECTIVE BODY LANGUAGE

"Your body is a hologram of your being, a three-dimensional movie that is constantly on, showing others how you feel about yourself and the world."
—Roman Polanski

OVERVIEW

Often a person will spend a great deal of time doing in-depth research, gathering facts, and preparing elaborate visuals, but doesn't think about the importance of the delivery of this material. This chapter will help increase your awareness of your nonverbal messages and suggest techniques to make your gestures enhance the words you are saying, rather than diluting or detracting from your content. Your delivery and body language cannot improve the ideas in your presentation, but they will determine how well your ideas are received and remembered by the audience.

There are two parts to a speech: the content and the delivery. In scientific and technical fields, audiences have been conditioned to believe that if information is important, with some semblance of organization and logic, the speaker's delivery can be haphazard. Realistically, if two engineers or scientists present essentially the same message to the same audience, the one with the more congruent, effective body language, voice, and presence will be more successful in eliciting the desired response.

The breakthroughs in technology that were intended to make communication simpler and effortless have actually created the opposite effect. Audiences try to escape the overload of information and electronic visuals that scream for their attention. They ignore or filter out what is not immediately relevant to them. The communicator has to search for ways to break through these barriers.

The answer is *not* to scream louder or to overuse additional high-technology stimuli. Your message must be clear, concise, brief, and relevant. By developing your nonverbal skills, your most important message will be heard and understood.

PRESENTATIONS ARE PERFORMANCES!

Before you even open your mouth, you need to show your audience that you deserve their attention. Much of the time, communication takes place without speech, by your body language and presence. If people do not like what they see in a speaker's appearance, they will resist the words they hear. If your delivery is forceful, brief, and dramatic, the listener is more likely to receive your intended message, and you are more apt to get results.

Model communicators are concerned with how they present their material. They know their delivery can be a deciding factor in whether they are perceived as credible and whether others are willing to accept what they say.

Many people tell me they are excellent communicators in one-on-one situations; they have problems only when speaking in front of a group. That is because public speaking involves performing as much as it does presenting information. You need to have a *sense of theater*, a sense of style to get your message across to your employees, customers, peers, superiors, or subordinates.

Let's compare content to food, and delivery to the way it is served. Take a lowly hot dog on a bun that has been carefully arranged on Dresden china. The waiter deftly serves it to you and silently waits to do your bidding. Add candle-light, an exquisite white linen tablecloth, aromatic roses, and beautiful soft music playing in the background. It can be quite enjoyable.

Now have someone carefully prepare you a delicate Swiss chocolate soufflé. The waiter throws it on a paper plate and slides the plate across to you on a greasy plastic tablecloth with such a shove that the soufflé slops over the side. The waiter turns on some heavy metal rock, tosses you a broken plastic fork, and grunts, "Eat!" It can be rather unpleasant. Delivery makes a difference!

WHAT ARE YOUR GESTURES, MANNERISMS, AND POSTURE SAYING?

Communication is a transfer of ideas, emotions, and energy. That energy is demonstrated through your body language, which includes eye contact, facial expressions, gestures, postures, and presence. Some people can energize a room simply by their entrance. Body language is the manifestation of internal feelings. It is usually more accurate than your words, which can be manipulated more successfully.

Although "body language" is a universal term, it really is a misnomer. To call something a language is to say that each component has three or four accepted meanings, and those meanings can be catalogued and listed in a dictionary. Gestures have a multitude of meanings, depending on the context, your cultural background, expectations, and previous experience.

Your audience receives over half of its information from your body language. Are your gestures adding to the meaning of what you're saying or are they telling the audience that you lack confidence, are nervous, or are unprepared? Before key decision makers give you control over people, resources, or money, you must convince them that you are indeed in control of yourself and your circumstances.

Most of us, when we're sitting in an audience, can pick up the nonverbal cues that alert us to the fact that a speaker is not entirely comfortable. Is he uncomfortable because he fears speaking in public or is he not prepared to intelligently discuss the subject? Nervous speakers have erratic gestures or none at all, have bland or tense facial expressions, clench their hands or play with objects, and may hunch over or slouch. Their eyes are unfocused and squinting, they shift their pace and their weight, or may appear devoid of energy. Confident speakers make no apologies, have a strong, confident posture, and a relaxed, alert manner. They calmly recover from mistakes, use few notes, and make good use of their personal space. They have animated faces, pleasant voices, strong eye contact, and warm smiles. What is your nonverbal language saying?

Going Beyond Yourself and the Material

There are three stages in the development of presenters:

- **1st stage** - involved with *themselves.* Concerned about how they look and how people are judging them. Does anyone want to listen to me? They are dealing with fears and anxieties.

- **2nd stage** - involved with their *material.* Is it accurate? Do I have enough statistics, proof, and visuals? Is it logical and well organized? Most technical professionals are in this stage.

- **3rd stage** - involved with their *audience.* Is the audience profiting from the message? Am I starting from where they are? Am I getting the response I want or should I modify my words or delivery?

Space engineer and science-fiction author Gentry Lee is an excellent example of a model communicator. He is involved with the audience from the moment he steps in front of them. He asks rhetorical and actual questions to get interaction and stimulate his audience. His gestures are spontaneous and punctuate his words. It is apparent that he is enjoying himself, and his contagious enthusiasm pulls in even the most resistant listener to discover what is so fascinating about science.

Your Body Language Reflects Your Mental Attitude

Several years ago, I attended a recital at the Northwest Suzuki School of Music. The children were having a group lesson and the headmaster gave a short critique after each child played. One child was having trouble with a series of fast notes. The headmaster told him that he was not playing the runs crisply enough. The child laboriously placed each finger on the piano keys. The headmaster stood up and solemnly declared, "You don't play the piano with your fingers, you play it with your mind!"

The headmaster was trying to teach the child that you will be more successful working on the *mental* causes of your errors than by dwelling exclusively on the physical. What is going on inside your mind? How do you feel about yourself, your material, the audience, and the situation?

> Thoughts determine emotions
> Emotions determine body language, energy, and vocal tone
> What are you thinking?

Anxieties, personal problems, health, and ego all affect your ability to influence your audience. Your inner emotions can conflict and destroy your real ability to sell your ideas.

Your body language grows out of your *self-image.* If you start to think you will be a failure, you may feel depressed about the whole situation, and your facial expressions and body language will reflect your uneasiness. If you have decided you will do a good job, your emotions will be upbeat and confident, and your enthusiasm will be evident in your posture, your eagerness to be understood, and the smile that is on your face when you begin.

All of us have days, of course, when we don't feel terrific about ourselves. You can ask yourself what would you do *if* you were confident, self-assured, and in control. Psychologist William James said, "Outward physical confidence, even though feigned, actually increases internal, psychological confidence." If you aren't feeling in top form, act as though you are, and your performance will convince everyone, including yourself.

Psychologist Paul Ekman, of the University of California, San Francisco, has shown that when individuals are asked to make faces corresponding to different emotions, their heartbeat, blood pressure, and other physical responses change to match the simulated emotions. For example, if you make faces corresponding to fear, anger, sadness, and disgust, actual chemical changes occur in your body. If you imagine you are happy and confident, your body's chemistry changes to correspond to those emotions. Your body language can actually influence what happens to you biologically.

What Kind of Animal Are You?

One client was having trouble running his weekly sales meetings and had been advised that he would lose his job if he could not take control. Although he was quite capable in many areas, he did not project an authoritative image. He seemed to recede into his short stature and was rather overweight. At our meeting I noticed a soft, pliable, dough-boy figure in the middle of his desk.

I asked the sales manager if he could think of an animal with characteristics that would help him reach his objective in meetings. He paused a moment and then declared, "An eagle. It has a commanding presence. It's majestic and bold. And it soars to great heights and can get perspective on situations." I suggested that he get rid of the squishy doll and replace it with his new image. Last Christmas his wife bought him a brass eagle with a three-foot wing spread that soars on his office wall. He reports that it reminds him of the strength he wants to convey and results in stronger body language.

Look around your workplace to see what your brain is constantly absorbing. What images are reinforcing themselves? What animal are you now and which would you like to be? Some answers have been revealing. One company president said that he was a snake because he liked to stay hidden, but wanted to be deadly when he struck. He wasn't interested in changing his image, but for his presentations I helped him communicate his qualities of alertness and cunning.

I asked a systems engineer to name an animal that had similar body language to his own. He answered, "A rabbit." Indeed, he rarely spoke and was timid. He decided that he should be something stronger, such as a dog, and his choice made a difference in how he carries himself. I suggested that he purchase a small statue of a Great Dane and put it on his desk. Now, each time he picks up the phone, he thinks of the characteristics of a powerful, self-assured dog, and it affects the quality of his voice.

Own Your Space

Model communicators are at ease with themselves and with their space. Their presence expands until it takes up the entire room, and they are clearly comfortable. Debbie Allen, one of the producers of the movie, *Amistad*, says that actor Morgan Freeman's rare, old-fashioned virtue is presence. "I have never seen anyone so powerful, even when he doesn't have anything to say. I watched him dominate scenes where he is just observing. He is kind of like a quiet storm, listening patiently, observing, understanding."

Here is an exercise to help you own your space: Pretend that you are very unsure of yourself. Hang your head. Act apologetic. Pull your body together so that you take up the smallest amount of space possible. Then stand tall, stretch your arms wide open, breathe deeply, and take up the entire room. Go back and forth from withdrawal to expansiveness. Feel the difference. The next time you walk to the front of the room to give a presentation, recapture that same feeling of owning the whole room. You can practice at your next office function.

Once you are visible to the audience, your body language can establish or destroy your credibility. Most speakers don't establish their presence until about five to ten minutes into the presentation; they may have already lost the audience. An actor realizes they have only a few moments on camera to make an impression. Olympic skiers may only have a fraction of a second to prove that they are gold medal contenders. Audiences judge a speaker *immediately*. You cannot afford the luxury of hoping you eventually create a good impression ten minutes into your speech. *Own your space and be involved with the audience the moment you step in front of them.* Take your time, pause. Let your audience become focused on you. Stand erect and breath deeply. Walk around (if it's possible and appropriate) to dissipate your tension. Smile, have fun, and enjoy the opportunity of having the attention of the audience. Let your attitude tell them that you would rather be here than anyplace else in the world.

Posture, Gestures, and Facial Expressions

Queen Victoria of England, who was only 5 feet 1 inch tall, often tied a garland of prickly holly under her neck to remind her to keep her chin up. Good body alignment conveys confidence and authority. Poor posture projects defensiveness, rigidity, awkwardness, and aggressiveness. During rehearsal and before you begin each presentation, go through a posture check.

Stand with your feet about six inches apart and with your weight balanced on the balls of your feet. Bend your knees slightly. Tuck your buttocks under and pull your waist up out of your hips. Shrug your shoulders and practice feeling comfortable with your hands down at your sides. Keep your chin parallel to the floor. Now, pretend to grow two inches taller. Imagine that your head, like that of a marionette, is being pulled up by a headstring that rises from your skull and goes up into the clouds.

If your body is in alignment, it will feel easy and natural for you to use your hands to make gestures. The old practice of walking around with a book on your head will force you to stand up straight. Try it! You will find that your hands fall comfortably to your sides. Avoid clasping your hands in front of or behind you, and don't fold your arms across your chest. These gestures keep the energy around you. You want your energy to flow to your audience. *Your audience will mirror your body language.* If you lean heavily on the lectern and talk in a monotone, your audience will go to sleep. If you are alert and energetic, your audience will be alert and energetic.

The futurist David Pearce Snyder uses strong, animated gestures that add to the meaning of his words. If you turn off the sound on one of his videos, you can see the vitality in his movement. If I turned off the sound on one of your presentations, would I be able to tell if you're enthusiastic about your subject? What are you saying with your nonverbal language?

Be careful about rocking back and forth or aimlessly wandering around. If you are using a pointer, remote control, or other tool or prop, put it down as soon as you have used it. An audience can become distracted when you play with toys. Much of this behavior is unconscious, but remember that small gestures can be distracting and often make you appear weak or nervous.

Your listeners receive a large percentage of information from your facial expressions. If your facial muscles never move or reflect what you are saying, are you enhancing your words? No one expects you to be an animated clown, but enthusiastic facial expressions can enrich meaning, even in a serious presentation.

Eye Contact

Your eyes will immediately signal your degree of confidence to everyone when you get up to give a presentation. Studies show that strong eye contact increases the audience's attention, interest, understanding, and satisfaction with the speaker.

You can either take in the participants and send out a message that says "we both count" or you can dismiss everyone in one sweep of your eyes. It is *not* a good idea to focus your attention on an object in the back of the room or on one or two people in the audience. Learn to have "private conversations" with the people throughout the audience. Acknowledge the eye contact of those who seem interested in your message and learn to grab and hold those who need to be brought into your presentation. Hold eye contact for a short duration; don't make someone uncomfortable by staring at him or her. Show the listeners that they are valued and accepted.

Author Ralph Waldo Emerson said, "The eyes of men converse as much as their tongues with the advantage that the ocular dialect needs no dictionary, but is understood the world over." Although we can mask our facial expressions, rarely can we mask what we say with our eyes.

If it is imperative that you read some lengthy statistics, be sure to glance up frequently and reestablish eye contact with your audience. Focused, direct contact will play a large part in establishing your authoritative presence in a room and increase the rapport with the audience.

CHARISMA IS FOCUSED ENERGY!

Your delivery should be crisp and energetic and reach out to the minds of the audience. Model communicators are enthusiastic about their work and feel a need to communicate their excitement to others. Author James Winans noted that, "An audience will forgive a speaker almost any lack if he is manifestly earnest about his proposal. Earnestness moves our emotions, thaws our indifference, and gives us faith, which a leader must create."

Astrophysicist and author Clifford Stoll is an excellent example of someone who combines an expressive voice and body language to convey the passion he feels for his subject. Stoll breaks every rule in the book for presenters and his audience does forgive him. He writes notes on his hand, precariously perches on the back of a chair, runs up and down the aisles, munches candy, and continuously gulps water. But his electric energy and involvement with the audience, along with fascinating content, mesmerizes his listeners.

Before you make a presentation, walk around briskly, shrug your shoulders, do knee bends, and feel the energy come from your toes. Let's see the enthusiasm in your facial expressions and hear it in your voice!

I often ask my clients to draw a stick figure of themselves and then draw arrows from the parts that radiate energy. Most people are puzzled and then draw an arrow from the eyes or hands. One woman said, "I never thought about energy below my neck." What parts of your body do you dislike? Some men and women are self-conscious about their weight or wrinkles or parts of their body, so they withdraw to protect themselves. You block your energy when you try to hide parts of your body. So start liking yourself, in spite of any shortcomings. Energy must come from all over the body, but particularly from the *solar plexus*. Pretend that a beacon of light emanates from your waistline. It will make an amazing difference!

If you are serious and steady and not given to shows of emotion, there is no need for you to be effervescent. One of the most charismatic people I ever met was a tiny nun who went about her work without noise or commotion, but who spoke and listened with intensity and concentration. Your style may be quiet and serious, but it is certainly salable. Radiate competency, personal warmth, and credibility from your entire body.

Speaking in public demands a certain amount of effort and produces some degree of tension. Relaxed alertness is impossible if you are physically exhausted. Flying coast to coast and dealing with jet lag presents a problem. If you are feeling fragmented, it will add to your nervousness. The body and voice must be rested. Get in training several days prior to your presentation by exercising, and eating and sleeping well.

Flexibility is important because you will be facing different audiences and trying to influence those audiences. It is difficult when you have spent time preparing your report and the person ahead of you speaks too long, people arrive late, or your computer crashes. If you are rested, it will be easier to adjust to inconveniences or distractions. You can focus on your main concern of getting into the minds of your listeners.

PRACTICING NONVERBAL COMMUNICATION

Spend some time observing other people's nonverbal communication. What do they do that's positive, that enhances their communication? Ask a mentor, a supervisor you respect, or a friend for objective feedback on your body language. Listen objectively, not defensively, and resolve to make whatever changes you can while being comfortable with who you are.

Model communicators say their delivery skills suffer most when they are inadequately prepared or not sure of their material. Obviously, your first consideration should be thorough preparation of your content. But do not neglect your delivery. In everyday conversation, sensitize yourself to the muscles in your face and feel what you are doing when you are expressing an emotion, qualifying a statement, or answering a difficult question. Do you squint or frown on a regular basis? Some people have a habit of looking angry no matter what emotion they are trying to express. Videotape yourself to find out if your image actually comes across to others the way you think it does. Eliminate annoying mannerisms such as jingling pocket change, pushing up your eyeglasses, or drumming your fingers on a lectern. Observe what is going on in the audience and capitalize on what is happening. Use the energy of the audience. Don't get caught up in being perfect. A creative, informative presentation needn't be flawless.

Frog or Prince?

Do you consider the people in your audience to be your superiors, subordinates, or peers? In other words, are they *frogs* or *princes*? Whether we like to admit it or not, we unconsciously make an immediate decision about our relationship to other people based on their authority, social status, educational status, material possessions, or physical appearance. That decision affects our body language and tone of voice.

Many speakers deny that they react to some people differently, but if you walked into a room of subordinates, wouldn't your body language and voice be different than if you discovered you were facing the chairman of the board and senior administrators? Monitor what changes occur in your body language when you attempt to persuade superiors or subordinates. Give others respect because of age, experience, or position, but talk to them as equals.

Effective Delivery Skills

In my experiences with scientists and technologists, it has been pleasant and refreshing to find exemplary logic and organization in their presentations. I salute them for their emphasis on substance. But I know they could increase their influence if they would pay more attention to the delivery of their information. You may have a better story than your competition, but you may lose a contract if they tell their story better. You will find that enthusiasm and openness to the audience will make you more persuasive.

Even if you don't have initial credibility, you can gain it during the presentation. Nonverbal signals can support and enhance the simplest statements. Believe in yourself and your ideas, and the value they have for other people. The audience will hear and see this. The unspoken word may have the most lasting impact of all!

KEY IDEAS

• Communicate your enthusiasm about your subject through your delivery.

• Be pleased with yourself. Your self-image is the most important factor affecting how you communicate.

• Spend time watching other people's body language, and get feedback on your own postures, gestures, and mannerisms.

• Make your body language congruent with your words and tone of voice.

• Own your space.

Chapter 15

DEVELOPING A COLORFUL, EXPRESSIVE VOICE

"For his voice could search the heart, and that was his gift and his strength."
—Stephen Vincent Benet
The Devil and Daniel Webster

OVERVIEW

The voice is an instrument to be used by the speaker to convey the total meaning of a message. You should be able to control and use your voice adequately, but always in an unobtrusive way. The effectiveness of your voice will be a critical factor in whether your opinions, insights, and instructions will be listened to and acted upon. This chapter will give you techniques for developing an expressive voice to successfully enhance your image and to effectively convey your information.

Pat Moneymaker, commander of the U.S. Navy's Blue Angels team, told me:

> I say the exact same 4,000 words every time we fly. There is no time to embellish the commands or comment, "It's bumpy out here," so I change the inflection in my voice to indicate the conditions, anticipation, and degree. For instance I may say, "addd powerrr" and stretch out the words. It's almost a second language for my pilots, and they understand if I mean to add power faster or slower. They only hear my voice, and they have to have absolute trust in what I am saying.

You may not be giving life and death commands in your next presentation, but your voice will either add to the understanding of your material or dilute your authority and your ability to persuade others. It will have a pronounced effect, good or bad, on your audience.

On several occasions, companies have asked me to work with their engineers or scientists on presentation skills, saying, "You can skip the instruction on voice. They don't need that."

What they don't realize, however, is that a *third of the information* that your audience receives will be from your voice. A clear, pleasing, expressive voice with precise articulation can accurately communicate your ideas and keep your listeners' attention. They will remember what you say and respond in the way you want.

I was fortunate to be a student of the legendary voice-over actor Daws Butler, who created the voices of Yogi Bear, Captain Crunch, Snaggle Puss, and more than fifty other cartoon characters. Three of us met for class at Daws' home. Usually we rehearsed scripts, perfected our timing, or worked on dialects. But one evening, Daws asked us to be seated, and for three hours he played selections of beautiful instrumental music. At first, I questioned what he was doing, but gradually I gave in to the various sounds and tempos, and noticed how they elicited different emotions. At the end of class, Daws summed up the lesson, "*Never forget your voice is a powerful instrument. Use it well.* Good night."

Voice makes the image whole. Other people define who we are by our voice and the words we use. Feelings, attitudes, physical state, and self-image are revealed by your body language and voice. Your voice should accurately reveal your emotions. Develop the instrument of your voice so that you are able to communicate the many fascinating dimensions of your personality.

ANALYZING YOUR VOICE

Have you ever felt that your audience isn't really listening, even though your information should capture their attention? Does your voice detract form the perception that you are an authority on your subject? Can your voice energize the audience? Do people frequently interrupt you or ask you to repeat your words? Does your voice add to the meaning of your words? You can increase the emotional and mental involvement of your listeners by having more resonance, variety, warmth, and vitality in your voice.

We acquired our voices by hearing and imitating family and relatives as we grew up. Sometimes we rebel against a loud voice that we've heard and adopt a very soft voice. But in any case, *the voice we use on a daily basis is not always our natural voice.* Fears and anxieties, tensions, and negative habits distort and prevent us from using our natural voices. The best way to produce a strong, pleasant voice is to allow the body to do what it naturally and efficiently wants to do. Voice production should be effortless, and your vocal chords should not become strained or fatigued.

To improve your voice, you must refine your listening skills to know when you're speaking correctly. It can be a surprise to hear yourself in a taped conversation. You may discover that your voice is higher pitched than you thought, or you may find you talk too rapidly or slur your words.

If you want to tone and firm your body, you outline a long-term strategy of diet, weight lifting, and exercise. Whereas you might also sign up for a face-lift or a tummy tuck, voices will only respond to hard work and practice. It is necessary to analyze your voice to determine which qualities are working well for you now, and to devise a plan to eliminate or minimize flaws. By setting goals, you can train your voice to function efficiently and well. A fantastic body is not an overnight achievement. Developing a compelling voice will take time, self-discipline, and practice.

DIAPHRAGMATIC BREATHING

Do not underestimate the value of proper breathing. Almost every problem of the voice—harshness, shrillness, a pitch that is too high or too low, choppy or annoying speech patterns—can be traced to improper breath control. Your speaking style, rate, quality, and resonance are affected by tension, which affects your breathing. The muscles of breathing should be exercised in the same way that we lift weights to exercise other muscles in our body, or stretch to become flexible.

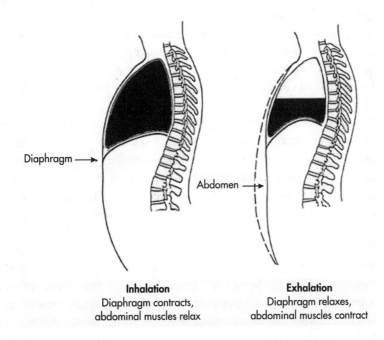

Inhalation
Diaphragm contracts,
abdominal muscles relax

Exhalation
Diaphragm relaxes,
abdominal muscles contract

The diaphragm is a dome-shaped muscle that divides the chest cavity from the abdominal cavity, and it is the single most important element in voice production. When air is inhaled, the diaphragm contracts, lowers, and flattens. This action creates a vacuum, and air rushes into the lungs to restore the balance. You may think that your throat is the primary source of your voice, but actually you should avoid using your throat as much as possible.

Join your hands over your stomach above your waist. Find the diaphragm by saying "whoa!" or "halt!" To feel your diaphragm moving, pant like a dog twenty times.

Try this exercise: extend your arms toward the corners of the room. Take a deep breath and pretend your lungs are a balloon you are filling from the bottom up. Then slowly, evenly, exhale. Release all the air until the balloon becomes flat and empty. Make sure you exhale every ounce of air. Keep relaxing all your muscles. *Do not allow your upper chest or shoulders to move when you inhale or exhale.* Feel your abdominal muscles, lower ribs, the rib muscles on the side, and back muscles expand slightly upon inhalation, like a bellows, and *relax* upon exhalation. Practice until it feels comfortable and normal for you to breathe from the midsection. If you have trouble breathing diaphragmatically, lean over a table or bend over from the waist or lie down. You will automatically breathe correctly.

Sound is produced when we exhale. Air is sent from the lungs through the trachea to the vocal folds in the larynx, which vibrate at different speeds to produce a variety of pitches. This vibration is given its unique resonance by the chest and lower throat, and the nasal and sinus cavities. Sound is shaped by the hard and soft palates, teeth, jawbone, cheekbones, and nose. The consonants are formed by the lips, teeth, and tongue.

Imagine a long, white, plastic pipe that extends from the middle of your diaphragm through the middle of your lungs, and up through your throat. Tilt your head backward and open your mouth, so that the pipe can come out your mouth and head straight to the ceiling. But since you cannot walk around with your chin tilted upward all the time, you need to relax your chin so that it's parallel to the floor. Thus, we have to insert a white, plastic pipe elbow that curves and directs the airflow out of the mouth. What we want is an unrestricted, unimpeded stream of air and sound passing through that channel. It's that curve in our throat, however, where there is usually a traffic jam. Any muscle tension in the throat, chest, jaw, or tongue will adversely affect the sounds we produce. *Relaxation during exhalation should be consciously induced and controlled.*

Have you ever watched the drawing for the lottery on TV? The lottery machine keeps all the numbered balls bouncing in the air. When the machine is turned off, the balls fall. Picture that stream of air keeping your words bouncing into the audience effortlessly. Send the sound through a free, open channel.

An erect but relaxed posture is important when you speak. This is not so much a matter of relaxing the muscles, as releasing the joints. Attach an imaginary string to the top of your head. Imagine your head floating up from your body like a balloon. Your neck should be free from tension and your spine should lengthen. Keep your body in alignment.

A phony voice cannot truly express your feelings. It is not necessary to use one voice for everyday happenings and another voice for public speaking. You play many different roles, and your speaking style should be adaptable. If you were asked to explain engineering basics to a high school class, you might use a different speaking style than you would use when addressing an international association of your peers, but the sound of your voice should remain the same.

A seventy-four-year-old widow began studying with me when she discovered at her fiftieth college reunion that a fellow classmate had never liked her because of her high-pitched voice. Her greatest moment came when the maitre d' at her country club refused to reserve her a table for Saturday night because he did not recognize her voice and admonished her for pretending to be a member. Another client told me he had always been rather quiet and faded into the woodwork during meetings. His new assertive voice earned him greater respect from his colleagues, and his views were now accepted as authoritative.

THE FIVE CHARACTERISTICS OF VOICE

Rate

Rate refers to how *slow* or how *fast* you speak, and is unconsciously influenced by your temperament and personality characteristics. If you talk extremely fast, people may perceive you to be very nervous. If you talk slowly and hesitate, you may be thought of as timid or unsure of your material. Your rate is also affected by your attitude toward what you are saying and your purpose with respect to your audience.

Often a speaker has a tendency to race through his material. You'll be misunderstood if you talk too rapidly, especially if you are trying to convey complex information. Slow down to deliver technical data or statistics, particularly when your audience doesn't have reference material. If you're reading material, vary the pace; speak slowly and deliberately when you want to emphasize a point. This will give those words more importance. Slow down when you refer to information in your handouts. If you are too slow and ponderous, however, the audience will tune you out. Be careful not to ramble, for it will dilute your authority. *Put periods at the ends of your sentences.* Make a powerful statement, then leave it alone.

Creating pauses can provide a pleasant and interesting variety of tempo. The talented pianist, Vladimir Horowitz, remarked that he played the notes, but the beautiful music was *between* the notes. A presenter can communicate volumes to the audience with skillful pauses between the lines.

Pauses provide you with a means of emphasis, a kind of oral punctuation that points out the significant information you want the listener to remember. A meaningful pause will tell your audience to reflect on what you have just said or to listen carefully to what's coming next. You can alter the pace of your information by talking at normal speed and then slowing down and pausing before a critical point, to emphasize and draw attention to it.

Pauses can be powerful. You can ask rhetorical questions and then pause while the audience thinks about what you have asked. Pauses can set up a line of an anecdote or add the twist in a humorous story.

Most people are shocked to discover when they listen to a tape of themselves that they use fillers such as "um" or "you know" during their pauses. These vocal interferences make the speaker appear nervous and they diminish credibility. Ask a friend for feedback. If you have this habit, try to talk for sixty seconds without using fillers. Be silent at the end of each sentence. In your everyday speaking, consciously strive to avoid the use of such distracting vocal mannerisms.

Practice standing comfortably during a pause. It will add authority to your presence. Pause a moment before you begin to speak, and you will show that you are in control of the situation. You can pause when you conclude your presentation and let the audience digest your final words. Stand there for a moment and then return to your seat.

Volume

Volume is the *loudness* or *softness* of sounds. The same volume, whether loud or soft, can become monotonous. Don't overuse loudness by "punching" words to emphasize your points; decreasing volume may also lend intensity to a special point. Maintain adequate breath control so that your sentence endings don't fade away.

Adjust your volume to the size of your audience, the physical characteristics of the room, and your proximity to the microphone. Arrive early at your speech site to check out the best volume level, but realize that when the room fills up with people, you may need to speak louder.

A Vietnamese engineer in one of my classes was very shy and barely spoke above a whisper. He had difficulty putting more enthusiasm and volume into his voice. I had him come to the front of the room with other class members and pretend that he was selling fish at Seattle's famous Pike Place Market, where the merchants keep up a steady, boisterous patter to entice the crowd to their stalls. The engineer was transformed in an instant as he persuaded prospective customers to buy his fish. I told him to keep that same attitude and asked him to begin his presentation. The class applauded when he finished his lively project update. He told me later that he had once been a fish seller in Vietnam before he escaped to the United States.

Don't confuse volume with projection. *Projection is the art of directing your voice to a specific target.* Your first requirement for adequate projection is sufficient volume so that the sound will carry as far as the situation demands.

Second, you need the right mental attitude. Projection is a product not only of breath control, but also of the speaker's awareness of the audience. Reach out to the entire audience and share your message.

Pitch

Pitch refers to *how high or how low your voice registers on the musical scale.* One of the easiest ways to make your voice expressive is to add pitch variety. Tony Greske described singer Whitney Houston as having a voice that "could sweep along up there in the upper register like the Concorde crossing the moon." Pitch is determined by your attitude toward your audience, your material, and yourself. A good way to find out your natural pitch is to hum quietly and then say your name and birth date. That pitch is the middle of your natural range. From there, you can go both up and down the scale.

Inflection is a variation of tone and is illustrated on a piano keyboard. Sound engineers will tell you that middle C on a piano keyboard has 256 vibrations per second. No matter how heavily or lightly you touch middle C, you will not vary the number of these vibrations. Every other note also has a specific vibration rate. If you have only one tone of voice, your voice always has the same number of vibrations. It will be monotonous and tiresome to others.

If you play a melody of only three notes, it won't be very interesting. The same is true of your speech. Speak in rainbows of colored notes instead of a somber monotone. High-pitched voices are frequently shrill and thin and do not convey strength. If you want to lower the pitch of your voice, care must be taken to avoid strain. Do not force the voice down. Merely let it drop to its natural landing place. Support the voice by breathing from the diaphragm. The pitch level will gradually become lower. Relax the throat. Think low.

Here is a good exercise to help you find your low-pitched, *natural* voice. Gradually bend over from the waist toward the floor and say, "The rock fell down, down, down." When you are in this position, you are forced to breathe correctly. You will find that switching to diaphragmatic breathing will lower your pitch.

Develop pitch variety by reading your morning newspaper and making extreme changes up and down your vocal range. I suggested to one engineer that she read fairy tales to her four-year-old daughter. The child won't stay interested in the "wicked, old witch" and the "fire-breathing dragon" unless the mother uses variety in her voice.

Develop a physical responsiveness to words. Say the word *soft*. Now visualize soft fluffy clouds, soft puffy pillows, and soft white cotton. Now say the word *soft*. One of my clients, who was describing research on wind tunnels, was able to picture them vividly and add variety to his pitch as he described the variables and results of his work.

Daws Butler, my teacher, reminded his students to touch their pronouns "like dust on a moth's wings." "Inflections are like jalapeno peppers," broadcast consultant Joanne Stevens cautions, "a little goes a long way." It is amazing what a slight inflection on the pronouns of "you," "we," and "us" will add to the intimacy of the talk and to your ability to draw in the audience. Say, "We want *you* to be satisfied with our service," or "I'm glad that *we* decided to sit down and talk about the problem."

If you end your sentences with an upward inflection, your audience will not feel that you have confidence in your statements. Ask, "Are there any questions?" while thinking, "I hope not!" Now ask, "Are there any questions?" while thinking, "I certainly hope so!" Notice the difference in the vocal pitch.

A monotonous voice will be perceived as less credible. Exercise your voice by reading some of your material aloud. Visualize someone you're always happy to converse with. See what happens to the expressiveness in your voice! By visualizing what you're saying and by using vivid language, you'll help the audiences form a similar image in their minds. Your reality will become their reality, and you will be more persuasive.

If you are speaking for forty-five minutes and have a 10,000-word speech, all 10,000 words do not deserve equal emphasis. Your audience cannot possibly remember every one of them. Emphasize important thoughts by putting them in *italics*. Imagine you have a yellow highlighter in your voice to create a vocal signal that the information is critical. Pretend you're a tour director, and direct your audience to certain key attractions. Be in control, but be lively.

A voice lacking in pitch variety often reveals tension in the speaker. The vocal exercises at the end of this chapter will help you achieve relaxed alertness.

Quality

Quality is the distinctive sound of your voice. It enables someone else to identify you merely by hearing your "hello" on the telephone. The quality of your voice is affected by your emotional state: happiness, relief, stress, fatigue, and anxiety are all apparent. Since voice quality is closely associated with mood and feeling, it will be influenced by your attitude toward your material, your empathy with the audience, and your desire to communicate. The audience can tell from your first words if you are confident and believe in your information.

In my classes, I ask clients to give a five-minute talk about a favorite cause, such as better education, abuses of the environment, or the use of animals in research. Invariably, their voices have color, expressiveness, and variety in pitch. Then I ask them to talk about a work project and their voices usually flatten into a monotone. Sometimes there is a "Well, you probably already know this" attitude. There is a direct relationship between their enthusiasm for a subject and the quality of their voices.

Theater actors must memorize two hours of dialogue, and then repeat the same lines for every matinee and evening performance as if they had never been spoken before. It is difficult to keep that freshness, that spontaneity in their voices night after night, month after month, and, for some actors, year after year.

The Mousetrap has been playing on the London stage for twenty-nine years with some actors from the original cast. The actors know the murderer's identity, they know the lines of every other actor, and yet they must react in believable ways. How does an actor do it? How does a speaker do it? *By concentrating and focusing all their energy on the present moment.* Internalize your ideas. The quality of your voice will be governed by what is going on in your mind. Lift the words out of your notes and make them come alive. Think the thought before you say it. Smile with the mind before you use your mouth, and your voice will smile.

Although you want a confident tone of voice, it must not sound arrogant. While an audience can truly admire intelligence, if it is wrapped in arrogance, they will be suspicious and uncomfortable. One of my friends has a superior tone of voice even when he talks about such mundane things as the weather. I am sure that his audiences must feel intimidated when he speaks about his aerospace projects. On the other hand, a young man from my service provider solved an e-mail problem over the phone, and had such a reassuring voice that I didn't feel stupid asking a basic question. Your voice should sound sincere and balanced, and should add to the overall impression that after carefully examining all the facts, you have arrived at well-founded conclusions.

Here's a technique I used with a female client who wasn't coming across as persuasive during her proposal presentation rehearsals. I asked her to recall the sense memory of a time when she was trying to convince someone about a cause that she believed in passionately. Then I had her imagine that she was an attorney defending her client on a murder charge; this client proposal was the final statement to the judge and jury. Her body language immediately changed as well as the quality of her voice. Believe in what you are saying and try this exercise of speaking "as if" you were a person of authority.

The quality of your voice is also determined by its resonance. As air passes through your chest, throat and head, it can acquire richness and fullness. Imagine your body vibrating like a tuning fork as sounds reverberate through your body. If you string a piece of catgut between two chairs and draw a violin bow across it, what kind of sound would you get? You would hear a very high and squeaky sound. But put that string on a Stradivarius violin and draw a bow across it, and you will produce a wonderful, rich, mellow tone. Blowing through the mouthpiece of a French horn produces a high-pitched sound. Put it back on the instrument and you will have a rich, mellow sound. In the same way, resonance cavities in your body contribute to the quality of the sounds you make. Don't be a talking head. Talk from your toes. Breathing exercises will help.

Articulation

Speech must be not only audible, but *articulate, distinct, and accurate.* Your listeners cannot be attentive if they constantly have to guess at your words. Remember that English may be a second language for many of those in your workplace, at association meetings, on videoconferences, and especially if you are speaking internationally. It is even difficult for southerners and northerners to "translate" each other's language. If your information is technical and complex, it is even more critical that your listeners be able to easily follow your words.

Every speaker should be careful not to omit endings or otherwise mispronounce words. An actor will purposely drop endings from words or substitute "jest" for "just" and "fer" for "far" to indicate a poorly educated person. If you say "comin'," "goin'," "wanna," or "tryin'," it will detract from an image of professionalism. Articulation exercises will help improve the clarity of your voice. Reading tongue twisters out loud as fast as you can will loosen stiff jaws, lazy lips, and a sleepy tongue. I've included some exercises along with a voice analysis checklist.

Body Warm-up Exercises

1. Yawn deeply and fully several times; sigh with a fully vocalized "ah."

2. Stand as tall as you can; stretch your arms over your head; reach up with one hand and then the other.

3. Bend over gradually until your hands touch the floor.

4. Bend your knees, squat, lower your head, and relax.

5. Slowly stand up, rolling up your spine, vertebra by vertebra, until you are standing.

6. Roll your head side to side. Rubbermouth your face: purse your lips and then pull them back into a smile.

7. Shrug your shoulders, shake your hands and wrists, and shake all over like a rag doll.

Vocal Warm-up Exercises

1. Hum through your nose. Try to feel the vibration on the inside of your head; bring it around to the sides; bring it around to the front. Create a buzzing hum around your head like a helmet. Feel the vibration on the bridge of your nose and sides of the head. Open your mouth slowly and go into an "ah" sound. Do this several times.

2. Place one hand on your diaphragm and the other on your back. Say the following phrases slowly, exaggerating the vowel sounds and running the words together. Make the tones rich, round, and full. You should feel your hands vibrate.

   • In her tomb by the sounding sea.

- A time to love, and a time to hate.

- One by one, they went away.

3. Say the consonants precisely. Use the front of the mouth and lips, teeth, and tongue. b, c, d, f, g, h, j, k, l, m, n, p, q, r, s, t, v, w, y, z.

4. Relax your lips. Say: buh bay, buh bee, buh bye, buh bow, buh boo, buh say, buh see, buh sigh, buh sew, buh fay, buh fee, buh dee, buh dough, buh do.

5. Relax the jaw. Repeat "chew-chaw" ten times. Pronounce the following words: choose, chip, chill, cheap, latch, pitch, peach, scratch, shop, tot, guard, not, hot.

Tongue Twisters for Precise Articulation

1. She stood at the door of Burgess' Fish Sauce shop, inexplicably mimicking him.

2. Big brown bumblebees were buried beside the bulbs in Bobby Brook's bulb bowls, baskets, and boxes.

3. The duke dragged the dizzy dragon down into the deep, dark, dank dungeon.

4. Matthew Mather's mother munches mashed marmalade muffins muttering about a multitude of misguided memories.

5. Rich rajahs ride reindeer with rakish red rope reins around their regal necks.

6. Six silky slithering snakes slid surreptitiously along, simpering and slyly sneezing.

7. Great crushing crates create great crumbling craters.

8. Terrific Tilda with thin, tawny hair tumbling to her toes was a tremendously talented typist for Teeper, Teeder, and Teeber on Tuesdays.

9. Lazy lizards sizzle in a drizzle; prize lizards are wizards with scissors.

10. Victor vowed vengeance and the valuable velvet vanished from the verandah.

SUGGESTIONS FOR IMPROVING YOUR VOICE

A tape recorder is indispensable for voice improvement. Record every presentation. Listen to the playback and work on one characteristic at a time. Identify your strengths and pinpoint areas that need to be improved. This time, you

VOICE ANALYSIS CHECKLIST

Rate
1. Could be slower _____
2. Could be faster _____
3. Can use more variety _____
4. Phrasing could be improved _____
5. Choppy, need to smooth out _____
6. Too many hesitations without meaning _____

Volume
1. Could be more expansive and full _____
2. Could be softer _____
3. Need more variety _____
4. Force (punch) overused as a form of emphasis _____
5. Need to project more directly to listeners _____

Pitch
1. General level could be lower, higher _____
2. Need more variety—sameness evident _____
3. Need to eliminate repetitive pitch pattern _____

Quality
1. Too flat—need more richness and resonance _____
2. Too harsh _____
3. Need to relax throat—strained, strident, shrill _____
4. Too timid _____
5. Need more variety _____

Articulation
1. Articulation of consonants needs to be more distinct _____
2. Endings of words omitted or slurred _____
3. Lips, tongue, jaw can be used more effectively _____

might want to work on eliminating fillers and slowing down your rate of speech. Next time, concentrate on increasing the variety in your pitch. Illuminate every shade of meaning in the words.

Read poetry or good literature out loud and make it emotionally expressive. Experiment with different vocal qualities or rates of delivery. Repeat the consonants (b, c, d, f, g) emphasizing different emotions. Be happy, pleading, seductive, conspiratorial, angry, sad, domineering, playful, or threatening.

Your personal life style will affect your voice. Drinking alcohol or caffeinated drinks dehydrates the body and can alter your voice. Irregular sleep and inadequate exercise will cause fatigue, which will be evident in your voice. Emotional stress will prevent your voice from sounding balanced and in control. Shouting or straining your voice can produce hoarseness. Sipping honey and lemon in warm water is soothing, or add a teaspoon of salt to a half cup of hot water and gargle a full five minutes.

Think of the range of notes, rhythms, and intensities in intriguing music. Work on the vocal exercises to help make your voice a powerful instrument. A compelling voice will help you present a more forceful image. You will enjoy hearing yourself speak and so will others!

KEY IDEAS

- Play your voice like an instrument.

- Use variety in rate, volume, pitch, and quality to keep your audience attentive.

- Talk from your toes and use your whole body to resonate or re-sound your voice.

- Demonstrate your confidence by emphasizing your thoughts with creative pauses.

- Use your voice to call attention to your message, not to the voice itself.

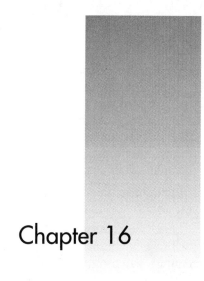

Chapter 16

USING HUMOR

*"Men will confess to treason, murder, arson, false teeth, or a wig.
How many of them will own up to a lack of humor?"*
—Frank Moore Colby

OVERVIEW

*Humor enhances your credibility because it shows you are confident and in
control. If you use humor appropriately, your audience will listen attentively and
positively; they will remember your message. Humor gives your ideas perspective and illustrates them succinctly. Although the use of humor can be risky,
speakers who choose relevant stories or one-liners that are appropriate to
specific audiences, topics, and situations can be very successful. This chapter
will show you how to incorporate humor into your presentation.*

Dick Lantz, a design engineer with NASA, began his presentation to a group of
engineers with a humorous anecdote using the familiar "rule of three," or "triad
device."

> A group of aerospace engineers were developing a new airplane
> design. During the first trial run, the wings came off—so it was
> back to the drawing board. At the beginning of the second test,
> everyone was convinced the errors had been corrected. The
> plane started down the runway with a roar and both wings began
> to vibrate and snapped off.
>
> The engineers were perplexed, but a janitor was watching the
> tests and offered a suggestion. "Drill a hole every 2 inches along

the line where you join the wing to the plane," he said. The engineers followed his advice, and on the third trial, the plane went down the runway, became airborne, and had a successful flight. The designer was flabbergasted. He went back to the janitor and asked him how he knew so much about stress concentration factors. "That's easy," the janitor said, "everyone knows that perforated paper never tears on the holes."

Lantz then stated his principle: Frequently in design work, it is easy to overlook the obvious.

People look forward to being amused. Humor in the form of personal stories or witticisms can ease tension, build better relationships, gain the audience's attention, revive interest, wipe out hostility, and help clarify your ideas.

Recently a client called me and lamented, "I'm not good at telling jokes but I know that everyone at the regional meeting will expect some humor. The head of our department always starts out with some hilarious story, and then I get up and am serious and boring. What can I do?"

Here's what I told my client: You don't have to imitate your boss's brand of humor. Humor isn't always a guffaw joke. Try a personal anecdote; it can be a powerful communication tool. A story that's appealing and reveals interesting information about the speaker can effectively create a bond with the audience. If the story elicits a chuckle or smile from an audience recognizing a truth or acknowledging a point, you've been successful.

When I ask company executives to name their favorite model communicators in science and technology, invariably they mention a speaker who uses humor. A person who successfully handles humor demonstrates to any audience that he or she is confident and in control. *It has been shown that an audience will be more receptive to a suggestion immediately after they have laughed.* Try getting your association members to laugh and then ask for volunteers. It works!

THE BENEFITS OF USING HUMOR

Some people think humor has no place in technical presentations because it detracts from the seriousness of a topic. The reverse is true: Because the topics are often dry, humor is a wonderful tool that can help a speaker gain and keep rapport with the audience. Many people confronted with technical information wonder if they will be able to understand and absorb the material. They may resist both the speaker and the subject. Laughter helps break down such obstacles, and makes the speaker seem more approachable and the material less forbidding.

When you laugh, both you and your audience let go of anger, frustration, and anxiety. T. George Harris, former editor of *Harvard Business Review* and *Psychology Today*, said, "An environment without humor invents nothing." Research suggests that putting listeners in a good mood helps them be more imaginative in solving problems. Humor is relaxing for both speaker and audience. Have you ever said, "I fell off my chair laughing!"? That is because laughter can release tension in your muscles. And remember, for the speaker, humor is one of the best remedies to counteract the fear of public speaking.

When I was attending an extremely technical videoconferencing seminar, a presenter from a nontechnical profession remarked that although he was pleased to be invited, he didn't know how much he could contribute to the meeting. "I feel like the farmer who entered his mule in the Kentucky Derby," he said. "He knew that his mule wouldn't win, but he felt the association would do it good." The audience laughed and relaxed. With those two sentences, the speaker established the parameters of his expertise and therefore prepared his audience. His humor had purpose, it was brief, and it won their attention.

Professor Marvin Minsky points out in *Society of Mind* that, "Laughter focuses attention on the present state of mind." If an audience laughs together, they become focused on the speaker and you have broken through the preoccupation barrier. Since the mind can have only one dominant thought at a time, if your audience laughs, you immediately know they are with you. This is also true with one-on-one communication. If your superior can laugh with you, a bond is established.

Many speakers can relate horror stories of presentations that went wrong, computers that refused to work, workers who started banging in the room next door, or a group of tourists who trooped into their session. One woman told me about a situation in a hotel conference room when the temperature dipped into the fifties and everyone sat huddled in coats. The handouts were lost between the mail room and the conference room. Blaring rock music was piped into the room at odd intervals. When management failed to respond to her pleas, the speaker moved the thirty attendees into the sunny lobby of the hotel next to a sign advertising the cocktail lounge. She quipped, "If this is what they mean by *Happy Hour*, can you imagine what we can expect later on this morning?" She got everyone to laugh about the inconveniences by developing a camaraderie with the audience and helping them to gain a different perspective on the situation.

SOME WORDS OF CAUTION

The flip side of attempted humor is that if it fails, you can lose your dignity. If humor is pointless and falls flat, it will be a definite setback and you will need to work twice as hard to establish your desired image. One engineer told me that the risk of standing up in front of everyone and looking like a fool is a definite deterrent to his use of humorous stories in technical presentations. Sometimes he attempts a one-liner, but nothing more ambitious. "It's a defense mechanism," he replied thoughtfully. "If I tell a joke or a story, I'm announcing my intention to be funny. It's easier to toss off a line or two. If no one laughs, I'm already on to something else, as if I hadn't really tried to be funny. I haven't invested time or my self-esteem."

He mentioned trying to explain that a processor his company was designing was too powerful for the client's needs. He thought it would strike a negative chord to simply say it was overkill, so he stated, "It's like driving a Ferrari in a thirty-five-mile-an-hour speed zone." It elicited a smile from everyone, and he was able to get a sensitive point across.

Are you comfortable with being humorous? Stay with the one-liners if you are successful with them. But in addition, try some personal stories. Audiences respond favorably to stories with a ring of truth. You need not abandon your own personality to be humorous. A subtle throw-away line can bring a smile. Art Buchwald made the observation, "Living in Seattle is like dating a beautiful woman who always has a cold." A sales rep quoted the cartoon *Dilbert,* "Never take an engineer on a sales call," he said, "he doesn't know how to lie." Many excellent speakers use cartoons or funny visuals to make a point when they don't feel comfortable telling stories.

Resist telling a joke merely for the sake of getting a laugh from your audience. Of course, there are always exceptions to the rule. When passing by a hotel banquet room where an Institute of Electrical and Electronics Engineers dinner meeting was taking place, I overheard the president's opening remark: "Somebody said I should be funny and tell a joke when I got up here." I winced inwardly but remained to listen, expecting to hear him fall flat on his face. But he told the joke with enthusiasm. It was hilarious and I started laughing with everyone else. The audience settled into a warm receptive mood for the president's rather dry material.

I have learned never to say "never." There are few things worse than an inappropriate story at an inopportune time. However, if you have a clever story that is tasteful and will survive under most circumstances, try it!

William D. Ruckleshaus, a model communicator who deftly uses humor in serious presentations, says that over the years, he has collected several thousand quotations and anecdotes. "They aren't really filed under special topics, but I have gone over them so many times that I know where to find the funny ones and the ones that will work for a specific occasion." After several presentations, you will learn which stories invariably succeed and which ones to eliminate.

A story that doesn't work at all for one audience may bring howls of laughter from another. Different cultures have different ideas about what is funny. If your comedy is dependent on subtleties of the English language and English is not the first language for most of your audience, you stand a good chance of failing with your humor. Inside jokes about your profession may be appropriate within your organization or association, but probably won't get a positive response from guests or the general public. If speaking to clients, read up on their current events and try a witty remark in reference to a hot issue. It would be wise to check it out first with your liaison.

Understand your audience and why they are present. If your listeners take themselves seriously, then perhaps you will gain more rapport by being serious and skipping attempts at humor. That same serious audience, however, might expect to be entertained during an after-dinner speech.

A pickup truck pulled up to the barn and the driver hailed the farmer. "How much is that old bull out by the road worth to you?" he asked. "Depends," drawled the farmer after a moment's hesitation. "Are you the tax assessor? Do you want to buy him? Or did you run him down with your truck?" As with humor, sometimes it just depends on your audience.

FINDING HUMOR IN ORDINARY OCCURRENCES AND OUR PROFESSIONS

Often, the ordinary things that happen to us in our daily lives tickle our funny bones. Our flaws make us laughable. Humor can be a painfully honest comment about ourselves, both individually and as a species. Garrison Keillor, the author of *Lake Wobegon Days*, and columnist Dave Barry are people who see the humor of everyday situations. The brothers Click and Klack, on the radio talk show, *Car Talk,* dispense advice about complex car mechanics in layman's terms, but lace their chatter with humor. It makes for an entertaining, as well as educational, show.

Comic Stephen Wright emphasizes:

> You must know how something works to find humor in it. We all know the normal angle at which you are supposed to see things. But there are loopholes in logic, because there are other logics. For instance, I went down to a store where the sign read: OPEN 24 HOURS, but the guy was closing. I said, "I thought you were open 24 hours." "Not in a row," he responded.

Wright's "logical" observations are made in a monotone voice:

> A lot of people are afraid of heights; I'm afraid of widths.
> If I melt dry ice, can I swim without getting wet?
> If the whole world's a stage, where is the audience sitting?
> Before they invented drawing boards, what did they go back to?
> How do I set my laser printer on stun?

Examine your profession for humor. It requires insight to joke about your specialty. Russell Morash, the director of Julia Child's TV cooking show, said that although Child's sense of humor is acclaimed, "it would be wrong to think of her as a clown. The woman is an incredible scholar and understands her subject so completely that she's able to impose humor on top of it."

One lab manager was exasperated with the indecipherable scribbling of instructions for lab experiments. She didn't want to be responsible for any disastrous results. During a staff meeting, she decided to try a humorous approach and remarked:

I feel akin to the TV executive who sent out the following memo:
To all concerned:
Please eliminate Roman numerals on any cue cards!
Undoubtedly, this is why the actor read the line
"And now, a few nostalgic songs from World War Eleven."

There are lots of industry-specific resources for humor on the World Wide Web. Check out the Ig Nobel Prize home page which reports a good-natured spoof of science and the Nobel Prizes. The ceremony honors people whose achievements "cannot or should not be reproduced." This year's winners included an award to the biologist who measured people's brainwaves while they chewed different flavors of gum. Another Ig Nobel Prize went to the researchers who discovered that listening to elevator Muzak stimulates immuno-globulin A (IgA) production, and thus may help prevent the common cold. A well-deserved award in Meteorology went to the gentleman for the revealing report, "Chicken Plucking as Measure of Tornado Wind Speed." One of the highlights of the ceremony was the world premiere of a new mini-opera, *Il Kaboom Grosso,* about the big bang.

A function of humor is to shake up our perceptions, expose frailty and ambiguity, and ultimately assure ourselves that it's really okay just the way things are. People like the unusual and unexpected twist. When Neil Armstrong was asked on the anniversary of his moon walk how the astronauts' lives had been changed by their adventure, he responded, "Before 1969, press confer-ences were much smaller."

Steve Allen has said, *"Tragedy plus time equals comedy."* You can remind your audience that strangely enough, perspective can give a humorous edge to their current negative situation. Several years ago when my mother died, the undertaker asked my sister to send down clothing and personal effects to dress my mother's body. Overcome with grief, I nervously walked into the funeral home and stopped abruptly when I saw my mother lying in the casket. "What is she doing with her glasses on?" I irately confronted the undertaker. "Agnes always wore glasses," he quietly responded. "Take them off," I demanded, "she can't see!" "But," he said defensively, "no one will recognize her without them." "I *don't* think it'll be a problem," I said, "Mom's the only one lying down!" My mother would have laughed at the absurd conversation, as I do now.

Contrast is the essence of comedy, to see incongruities in normal events. If we say that sense is the normal order of things, then incongruity makes *non-sense.* For example, Albert Einstein played the violin and was once playing in a string quartet with Gregor Piatigorsky, William Primrose, and Jascha Heifetz. Einstein was not in the same class as these masters, but was allowed to play second violin. During one passage, Einstein repeatedly came in at the wrong time. Finally, an exasperated Piatigorsky stopped and blurted out, "Mr. Einstein, can't you count?!"

The prolific science writer and author Isaac Asimov explained, "It is the quick flip from sense to no-sense that is itself the change in point of view and that brings the laugh." Asimov related the following paradoxical story on television:

> On the glorious day of July 20, 1969, when the first human being stepped onto the surface of the moon, an Israeli said to an American friend, "That is a great achievement, but we Israelis will do much better. We are planning a manned expedition to the sun."
>
> "To the sun?" exclaimed the astonished American. "But the heat? The light? The radiation?"
>
> The Israeli chuckled. "Do you think we Israelis are fools? We will send the expedition at night."[1]

The key to creativity is looking at one thing and seeing another. Comedian Emo Phillips mused, "I used to think that my brain was the most important organ in my body, but then I thought: Look who's telling me that!" Watch for the inconsistencies in everyday logic. It could be an observation made by your child or the strange way the waitress served you in a coffee shop. Can you relate some amusing or clever everyday happening to a point in your speech?

A telephone company spokesperson began his talk to a public gathering about a rate increase by saying, "It's true that long distance rates may go up. That's the bad news. The good news is that the continents are drifting closer together." The potentially hostile audience was so surprised by the unexpected humor that they laughed. Then the spokesperson acknowledged the audience's concern about the rate increases so that they knew he wasn't being flippant about the situation. When emotions are running high, it is mandatory to lighten things up if you want your message to be heard at all. If done well, humor can show that you aren't rigid when everyone else is uptight.

KNOWING THE BEAT

Timing is a critical factor when telling a story. Timing dictates how long you pause before, between, or after words or sentences. It is knowing when to pounce on words, to hold them, to start again, to throw them away, to aim them precisely. Bob Hope has said that, "Perfect timing is the meshing of your brain with the audience's brain."

Several years ago, when I was playing opposite Shelley Berman in the play *Don't Drink the Water*, Shelley told me to walk on stage, say the first part of my line, set down my suitcase, finish the line, and then look at him. I wondered how he could dissect the line so minutely. I decided instead to do it my own way. The result failed to provoke even a chuckle from the audience.

The next night I did as Shelley told me. The result was uproarious laughter. He gave me an "I told you so" look. Setting down the suitcase provided a pause that added an extra beat to the rhythm of the line.

Timing also includes the use of body language. We usually pay attention to the delivery of our words in a story, but the audience gets most of their information from our gestures, our facial expressions, the inflection of our voice, and what *isn't* said. This is why it's necessary to be *totally involved* when you tell a story. Don't be tentative; jump in with both feet!

If you have a funny story that hasn't worked well in the past, think of it in terms of a song. You may have to add another word to make the beat of the line come out right. Adding a pause, a look, or a gesture before the punch line can highlight the words that follow. Watch and listen to your favorite comedians and comediennes. Observe their body language, their total involvement in what they are saying, and, especially, their timing.

Timing also means being prepared to change or eliminate your story depending on the circumstances. One speaker was adept at incorporating humor into the first few moments of his speech. However, prior to his introduction to a local organization, the president called for a moment of silence in memory of a member who had died that week. The speaker knew that he couldn't possibly begin with a joke, and it was not until the middle of his presentation that he felt he could introduce a humorous tone.

HUMOR IS AN ATTITUDE

We are all born with the potential for being amusing. If you have suppressed your sense of humor as an adult, it can be revived. Humor is more than telling a joke; it is the ability to be delighted with life. Do not allow a poor response to your attempt at humor keep you from trying again. Humor is subjective, and if it doesn't go over well the first time, it's not always a reflection on you or your choice of stories. Often there are other factors affecting the audience's response. It can be difficult to "read" technical and scientific audiences. One presenter said, "Your timing and confidence can go haywire if you assume from the silence that your first story is a dud. I've found some audiences may be laughing inside while displaying a stoic attitude outside."

Here are some tips on how to make humor work:

- *Analyze your audience.* What will appeal to them? What image will work best for you in this situation to obtain your objective? Would cartoons, animation, or funny visuals emphasize your points and be appropriate?

- *Make your story relevant* to your message.

- *Be clean and lean.* Write out your story and edit it to take out the "padding," the extra words and unnecessary details. It shouldn't be longer than a minute unless the point it makes warrants taking time from the main body of your speech.

- *Provide a transition* from your story to your topic and blend it in. Make the story sound as spontaneous as possible.

- *Avoid off-color stories* or jokes about minorities, women, the disabled, or ethnic groups. Remember, the purpose of humor is to enhance your message.

- *Display your modesty and humility* by telling a story that reveals an embarrassment or a minor failing. This allows the audience to feel superior. But don't humble yourself to excess. Such stories and put-downs can work only if the speaker is clearly very confident.

- *Use the "rule of three."* Give three examples or three times a person did something.

- *Use descriptive words* to call up vivid images. A humorous story is like a balloon: pump it up with details and then puncture it with a punch line.

- *Make sure everyone hears the punch line*, but don't dilute it by explaining its meaning.

- *Be precise.* There are no uh's, if's, or maybe's in humor.

- *Summarize your message with a witty or humorous quote* and leave the audience with a warm feeling.

- *Be totally involved in the story as you tell it.* Believe in it! Take risks.

- *Rehearse! Rehearse!* Repeat it over and over again to anyone who will listen so that you are comfortable with the words. Practice telling a story under similar conditions that you will face, in front of people who are representative of your audience. (This doesn't mean your dog.) The late actor Don Ameche related a story to me about the legendary comic Jack Benny, who accompanied him to a charity luncheon in Hollywood. Benny knew that someone would recognize him, and he spent an hour trying out several surprised looks and quips for Ameche's approval. Later, at the luncheon, Benny was recognized and introduced. He feigned amazement, as he rose to applause. Ameche said that Benny used the exact ad-libs and body language that he had rehearsed.

- *Start keeping a notebook or file of amusing quotations, cartoons, news stories,* and *personal stories* and arrange them topically. Use joke books and anthologies to *suggest* ideas. Clip and save amusing quotations, funny newspaper stories, cartoons, or stories from magazines. Check out the Internet for special Web pages on humor. Good stories sweep the country by fax and Internet, and can lose their punch because everyone has heard them. On the other hand, a personal anecdote is uniquely yours and will be fresh to the audience.

When you have the substance of your presentation's main points and supporting points outlined and written down, review your material. How can humor support your points? How can you use humor to illustrate, give perspective, or clarify images? Can you add a personal story, anecdote, or humorous quotation or plan a quick ad-lib? Humor will strengthen any presentation if it is relevant, appropriate, tasteful, and properly timed.

KEY IDEAS

- Use humor in your presentation to add to your credibility.

- Share personal stories to create a bond between you and the audience.

- Collect humorous anecdotes and quotations.

- Sharpen your delivery and timing by rehearsal.

- Cultivate your sense of humor by looking for the nonsense in daily life.

Notes

1. Excerpt from *Treasury of Humor* by Isaac Asimov. Copyright © *by Isaac Asimov.* Reprinted by permission of Houghton Mifflin Company. All rights reserved.

Chapter 17

HANDLING QUESTIONS WITH EASE

"A sudden, bold, and unexpected question doth
many times surprise a man and lay him open."
—Francis Bacon

OVERVIEW

The question and answer (Q&A) period should be thought of as an extension
of your presentation. Welcome this opportunity to clarify your ideas and fortify
your message. The audience will be able to see your mastery of the subject and
how well you think on your feet. This dialogue can also provide valuable feed-
back, with a clearer understanding of how the audience has examined and
accepted your ideas. This chapter gives you ideas on how to make your Q&A
session the most lively, thought-provoking, and successful part of your
presentation.

Few people invest in an expensive suit without trying it on and most people
need to try on your ideas before they accept them. Encourage your listeners to
participate in the Q&A period and involve them mentally and emotionally. If
they can question, explore, discuss, challenge, dispute, or probe your subject
matter and have their uncertainties satisfied, they will be more likely to accept
your ideas. Giving a monologue is passé, especially in a sales situation. Everyone wants, expects, and almost demands *interactivity*. If electronic visuals
dominate the presentation, the question and answer period assumes even more
importance for establishing rapport. Questions provide an opportunity to
strengthen your message and add to your persuasiveness.

Many times a question will reveal a controversial point that you neglected to cover in your presentation. A convincing answer may mean the difference between the audience's acceptance or rejection of your entire presentation. Realize there are no "canned" answers. Different audiences or situations will require different responses to the same question.

A client of mine was preparing to present a proposal to a prospective client. "I hope I don't get any questions," he said. "We have so many new features to our product that I'm not sure I could answer everything adequately." Prepare well-thought-out answers for expected questions; have a *plan* for addressing unexpected questions.

Decide how you can reinforce your message and advance your objective to inform, instruct, report, or persuade during the Q&A period. Anticipate which questions might be asked and prepare answers for the most difficult or controversial. Review your audience analysis checklist. Who are the people in your audience and what is their background? Can you expect them to accept your ideas or to challenge you? For each possible question, decide how you will finish and start, state your main point, support that point, and use the best organizational pattern. Wow your audience by having an extra viewgraph or slide, or be able to access computer-generated data that will provide the answer. Can you add variety to the Q&A session? If you anticipate controversial questions, rehearse with your peers and go through some "what if" scenarios.

An aerospace engineer told me that he spent weeks preparing for a presentation to a former Boeing chairman. He said there was only one question for which he had no answer, but he felt that he covered all of the other information so well that it might not matter. When he finished his presentation, the chairman asked him the one question he dreaded. And he wasn't able to answer it!

An engineering firm made the cut for a large government contract. The bid would be determined by a final Q&A session. A team of peers played the devil's advocate and grilled the two presenters on every possible question they might face. This rehearsal helped clarify which topics still needed additional facts and statistics, and eliminated surprises in the actual session.

TONE OF THE Q&A SESSION

Your attitude during the Q&A session can demonstrate your control and further your credibility. The audience will usually go along with any approach to the Q&A session as long as the speaker shows confidence and leadership. If the listeners sense you are weak, uncertain, capricious, or quarrelsome, they will consider their time wasted and resist taking part in the session. Select words that define your image; you might want to come across as calm, concerned, in control, or helpful. Remind yourself that your body language and tone of voice should reflect the image you have chosen.

One engineer, who was experienced in answering mediation questions or acting as an expert witness in a courtroom, said that he is usually part of a design team. "The team can include representatives from an architectural firm, a structural firm, a mechanical firm, and—depending on the site of work—a civil engineering firm. We meet beforehand to go over the specific claims so that we can put on a unified front. It would be folly to have disagreement within the team."

He continued to explain, "When you are in court, you have to be careful what you say and how you say it. It's not productive to be anything but calm or at least exhibit calmness, even though you may be really mad inside. If there's a claim that you believe is false, you want the opportunity to explain and defend your work."

If you ask for questions while standing behind a lectern, chances are that this barrier will discourage a response from the audience. Get out from behind tables or computer equipment. By being open and vulnerable, you will encourage your audience to risk a question or comment. Stand in close proximity to your audience, *raise your hand*, and ask, "Are there any comments?" Or you might ask, "Would anyone like to start?" or "I'd like to hear any remarks that can add to our discussion." Your vocal tone will indicate your attitude. Send your audience a message that says, "I hope you have questions!" or "Let's continue this dialogue!"

MOTIVES OF THE QUESTIONER

Pay special attention to the first question you hear, since several people may want that same point clarified. If there is a large audience, restate the question so that everyone can hear it and feel included in the discussion. Concentrate on the questioner and listen intently. The audience will sense immediately that you value what is being said. Try to catch the *intent* of the questioner, as well as the content of the query. You will find that the members of your audience usually ask questions for one of the following reasons:

- They are interested in obtaining additional information, a clarification, or an interpretation.

- They want your personal insight, recommendation, or judgment.

- They want to show you or the rest of the audience how smart they are.

- They want to embarrass or intimidate you. They may draw attention to what they consider to be incorrect statements, try to invalidate your thesis, challenge your sources, or attempt to anger you.

If you are making an in-house presentation and know the styles and motives of the principal players, you may be able to anticipate the types of questions (and questioners) and prepare accordingly.

Tips for Your Q&A Session

- *Reinforce and expand on your objective.*

  A question and answer period gives you a second chance to plant a message, and to expand and reinforce your point of view. When your audience leaves, what main idea do you want them to walk away with? To be persuasive, note how your plan, proposal, or request will produce a beneficial effect for each of the different members of the audience. Then, during the question and answer period, take every opportunity to build your case by outlining the benefits.

- *Think about the question before you think about the answer.*

  Time can be used to your advantage. A pause can give value to the question and give you enough time to search for the most concise way to answer the question. Don't overuse expressions such as, "Glad you asked that question," or the audience will know you are stalling.

- *Establish strong eye contact.*

  One researcher had a habit of looking down at the floor when she was asked a question. In reality, she was assessing the question and preparing her response, but her body language gave the impression that she didn't know the answer. Maintain direct eye contact with the questioner. If you are searching heavenward for inspiration, you will miss subtle body language and facial expressions that reveal *why* a person is asking and the best way to answer. You can also check the reaction to your response. Look at the rest of the audience as you repeat the question, and include them as you give your answer.

- *Watch filler words.*

  Think silently. Too many uh's, ah's, and you know's not only are distracting but will make you appear uninformed.

- *Be direct and concise.*

  Give a clear answer up front. You can expand your response if you feel that a longer explanation is necessary. The SAFW (Statement, Amplification, Few examples, and Windup) organizational pattern works well for answering questions concisely. Be wary of answering the unasked question.

- *Don't ramble.*

  Your objective is to answer this question and move on to the next one.

- *Use humor if appropriate.*

If a humorous answer will make a point, use it. Some of the best model communicators are known for their wit. Humor makes you appear relaxed and confident.

- *"Bookend" your answers.*

Make a direct answer at the beginning and, after expanding on the point, make it again at your conclusion. "So, as you can see, we have taken the necessary steps...."

- *Use rich imagery.*

The audience is able to visualize your answer much better if your language is vivid. When Neil Armstrong was questioned about the use of robots to explore space—a technique favored by many scientists—he replied, "Man can be amused and amazed, but a robot can be neither."

- *Choose several personal anecdotes and examples beforehand that can add interest during a Q&A period.*

They can rarely be made up on the spot. The late Governor Dixy Lee Ray told me she was frequently asked, "Why can't a nuclear power plant explode?" Her answer was, "Radioactive material can be compared to flour. Flour in its dry state can blow up. There have been explosions in silos and flour mills. Mix the flour with water and you form dough; you have changed the physical state. Dough doesn't explode. Add other ingredients and bake it and you produce a loaf of bread. Bread doesn't explode. It is the same with radioactive material—it exists in a different physical state in the power plant and can't explode."

- *Adapt your body language and tone of voice.*

Adjust your rate of speech to the questioner's speech, and mirror his or her body language. One client told me that he went toe-to-toe with a hostile questioner and then slowly relaxed his body. He lowered his voice and slowed his speech. He was amazed that the other person began to mirror his response and they parted as friends.

Avoiding Land Mines

- *Wait until the questioner finishes.*

Don't assume that you know what the person is asking and how you are going to answer before the question has been completed.

- *Clarify the question.*

With a large audience, repeat the question to make sure that everyone has heard it. Divide long or complex questions into parts. Admit it when you don't understand the question. If you aren't sure what the person is asking, you will be at risk for an inept answer. Ask for clarification, especially if the question can be interpreted in several ways. Listen carefully to general terms that crop up in questions and ask for specifics. "Who are 'they'?" "Can you give an example of what you mean by 'most objects'?" "'Better' as compared to …?"

- *Acknowledge the questioner's feelings.*

Never deny the person's feelings. Don't ever say, "I understand how you feel," unless you have been in the same position.

- *Avoid "yes" or "no" answers.*

Even if it is the correct answer, use the opportunity to elaborate or explain. Direct the question toward an issue you want to cover, or change the subject. "Yes, we do expect to reduce staff this year, but I want to point out…." "No, but it is more important to consider…."

- *Clear up assumptions in the question.*

Be careful when answering questions such as "Why are you raising your rates and overcharging us again?" If you don't challenge the assumptive language or the negative points in a question, some people will accept them as fact. Rick Chappell, associate director of science for NASA Marshall Space Flight Center, was asked, "Why are we spending money in space when we have the homeless to take care of?" Chappell answered, "Money isn't spent in space, it is spent here on earth." Then he went on to tell about the medical advances and benefits for the ordinary citizen from space research, such as generating the new technology that will keep America competitive and will strengthen the economy and, hence, our quality of life.

- *Watch out for unfamiliar statistics.*

If a questioner gives statistics unknown to you, ask for the source before giving your answer or opinion. You have as much right to question sources as the questioner does.

- *Listen for illogical reasoning.*

One woman confronted a representative from a company that was voluntarily cleaning up toxic waste. "I have constant headaches and have been ill all winter," she said. "People who drink contaminated water have headaches. Your plant is contaminating my drinking water!" The spokesman replied, "I can understand your concern about your health. Your headaches and other health problems may be attributed to other reasons. Our studies show that your drinking water is not contaminated at the present time. Our company will have the cleanup completed in eight months, and we will keep in touch until it is done satisfactorily."

- *Control irrelevant questions.*

If someone asks a question that has no real bearing on the subject, acknowledge the person and the concern, but diplomatically get the discussion back on the subject matter.

- *Admit when you don't know.*

There will be times when you are not able to come up with an appropriate answer. Tell your audience that you don't know in a strong, confident manner, and say, "I will get back to you with the information." And be sure you do! Another possibility would be to say, "I don't feel that I should comment on that because I don't have the latest test results. However, I can say..." and go on to make a brief, positive point. If appropriate, this might be a chance for you to facilitate the questioning by involving other members of the audience. If there is someone else present who has the expertise to answer the question, you can request his or her comment. Can you set this up beforehand? If you sense that the questioner wants to show off his knowledge, you can direct the comment back to him and ask how he sees the situation or what his information reveals.

This situation was handled well by a software representative at an association meeting. When an audience member asked a complicated question, the rep asked for the attendee's business card and e-mail address, jotted down the question on the back of the card, and said he would research the answer and contact him.

- *Find out if the media will be present.*

Know whom you can trust if there are media representatives present. Always remember there are no such things as "off-the-record" remarks!

- *Avoid questions you don't want to answer.*

You might respond, "That is an important consideration, but we feel our customers are more concerned about...," then bridge to something else.

- *Narrow down nonspecific questions.*

You may have to curb the participant who talks on and on without getting around to a question by politely interrupting and asking him or her, "Did you have a *specific* question?" You may also dissect one or two of the person's points that will interest the audience and expand on them.

• *Handle the person who tries to dominate the Q&A period firmly but courteously.*

Sometimes a participant won't let you finish your answer. Be polite. Ask to finish answering one question before going to another. Be in control. If it is obvious that one person is trying to monopolize the discussion, invite him to speak with you at a later time. Say that you want to give others a chance to get into the discussion, then quickly turn away and make eye contact with someone else who has a question. Never yield to the temptation to ridicule anyone.

One engineer cited an instance when a person on his review board demanded the engineer produce more data to back up his findings. The board member would not accept the answer that data was not immediately available. The engineer could see that the reviewer's motivation was to demonstrate his power in the group.

The engineer calmly replied that he would be happy to collect some substantiating data and deliver it in a day or two. If that was not satisfactory, he said, he would need additional resources to fulfill the request completely. At this point, another board member asked if it was worthwhile to spend extra money and time to find the data, and asked how the information would influence the decision they were supposed to make that day. The group decided the effect would be minimal. The engineer's controlled and cooperative manner of handling the questioning alerted the other board members to a power-play digression and they acted to bring the discussion back in line.

• *Acknowledge that others may hold contrary viewpoints.*

One scientist began his Q&A session by saying, "I don't expect everyone to agree with everything I have said here tonight, but I think that we are all in agreement that safety is the paramount consideration. I would like to hear your responses or questions in regard to my proposal." He reminded everyone of their common ground and that their opposition would have to be weighed against higher principles.

• *Avoid answering a hypothetical question that lures you into a negative scenario.*

When asked, "But what if you can't meet the production schedule?" you can confidently state, "We have completed this and this. The schedule is workable."

- *Own up to your previous statements.*

One of my clients had a sudden and complete turnabout in his beliefs on a certain subject. At his next presentation, a questioner produced an old newspaper clipping and demanded to know the reasons for the apparent change of heart. "Yes, those were my words a year ago," he replied, "but certain facts have been brought to my attention and they have convinced me to change my mind."

- *Be calm and objective in a hostile situation.*

Rather than firing back a denial or defensive response, pause, look directly at your adversary, think carefully about your reply, and then give a positive answer. You can depersonalize a question by rephrasing it in a neutral way; do not repeat a negative phrase. Your answer need not be an exact response to the query, as long as it is true to your interpretation of the question. Recognize that some people wish to hold on to their beliefs regardless of the facts and statistics you have to back up your contention.

- *Keep everybody on the same page.*

The technical language for the Q&A session must meet the needs of the entire audience. You may need to translate acronyms, define terms, or clarify exactly what a person is asking. You set the technical parameters for how complicated the questions will be. If there is a particularly complex question that would interest only a few members of the audience, invite the questioner to meet you afterward to discuss details. Don't waste audience time on answers that have no bearing on your objectives for the session.

No Questions Means No Applause

If there are no questions, pause a moment and then say, "One question I am often asked is…," which may nudge others into taking a risk. If you spoke with members of the audience prior to your presentation, you can say, "I was asked earlier why…" or, simply close with, "I would like to leave you with…" and restate your main idea or request.

If possible, you might interview members of the audience before your presentation and ask them about their major concerns and questions. You can pass out small cards and ask your audience to write down questions as they come into the room. This is particularly helpful if the queries will be technical ones that might require extra data or visual aids.

Videotaping or audiotaping your responses to questions can be revealing and will help you work toward a clear, crisp delivery.

Compliments

An audience can learn a great deal about a presenter by the way he handles compliments. General H. Norman Schwarzkopf, commander of Operation

Desert Storm in the Middle East following Iraq's 1990 invasion of Kuwait, was being interviewed on television. A reporter mentioned the colorful array of medals on the general's uniform. He asked, "Isn't that more medals than any man should carry around?" Schwarzkopf deftly responded, "This is a tribute to the poor marksmanship of the enemy."

Your Second Ending

Question and answer periods are commonly held at the end of a presentation. Unfortunately, they often divert attention from your intended message and final impact. Be wary of saying, "I will take one more question." It may be trivial, irrelevant, negative, or too complicated to answer in a short time, and the whole session could wind down unenthusiastically.

End on a positive note, before the audience tires or starts to leave. If you feel you have answered a question well, say, "That's all for now, but I will be glad to answer further questions privately after we conclude." Bring the focus back to you and summarize your main points in about 60 seconds. *This short summary after the question and answer session is a vital part of your speech.* Prepare it well. You may need to refine it slightly in view of the questions you've fielded, but get back to your main message. Help your audience to focus on what is important; they will remember most your *last words.*

KEY IDEAS

- Gear your Q&A session toward the objectives you want to accomplish.

- Prepare possible questions and practice your responses.

- Decipher the motives of the questioner and acknowledge his or her beliefs or feelings in your response.

- Use hostile questions as an opportunity to clarify your points.

- Summarize your main point at the end of the Q&A session and end on a high note.

Part IV

MANAGING YOUR CHANGING ENVIRONMENT

CHAPTERS

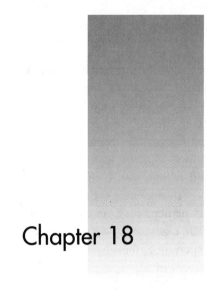

Chapter 18

SETTING THE STAGE

"The mediocre person is ruled by his environment, whereas the successful man uses the pressure of adversity as an assist in obtaining his final objective."
—Clifton Burke

OVERVIEW

This chapter gives ideas, insights, and practical solutions for controlling your environment so that the audience's attention is focused on you and your presentation. Creating an atmosphere conducive to your audience's listening, responding, and remembering will make your goals easier to achieve. Maintain your professional image and be ready to adapt your material and your style if the site of your presentation is less than perfect.

You have prepared well for this presentation and have even had time to rehearse with your visuals. You feel in tip-top shape, and your audience is enthusiastic to hear what you have to say. Unfortunately, you didn't anticipate the retirement party next door. The smell of fresh-brewed coffee, baked beans, and hot dogs wafts down the hallway, and you have to raise your voice to be heard over the noise, laughter, and steady background beat of "Rock Around the Clock." And when your computer-generated visuals shut down in the middle of the presentation because of an electrical overload from the party next door, it's time to postpone the meeting and join the fun.

The ill-fated engineer who endured this scenario was thoroughly prepared for his presentation, but had neglected to check out his environment in advance. It proved to be his downfall. If you have encountered problems getting the response you want from your audience, have you considered that the site may be partially responsible? Perhaps the reason people are distracted, irritable, or sleepy has something to do with the physical surroundings.

Julie Swor, member of the International Association of Conference Centers, asserts that, "The physical environment sets a psychological mood, and research has shown the tremendous effect of the total atmosphere on learning. Light, colors, climate, and physical comfort combine to affect the brain's concentration and receptivity to information."

An audience's first impression of a speaker can influence their acceptance of his or her ideas; so can the audience's first impression of the site strongly affect their reaction to what follows. Robert W. Lucky of Bellcore comments, "In my experience, the physical setting of the speech is one of the most important factors of all. You can give an identical talk in two different settings and get vastly different reactions. A lot has to do with the seating arrangements, time of day, etc." Every room has certain dynamics: there are friendly rooms, and rooms that work against the speaker. Windows, carpeting, acoustics, wall colors, tables, the placement of the lectern, seating, sight lines, and other factors can either add to the participants' comfort and responsiveness or distract them to the extent that you and your message are lost.

WHAT IS YOUR OBJECTIVE?

The environment should match the purpose of your meeting. If you are conducting a day-long training session, the seating arrangements will be different than for a forty-five-minute project update, a panel discussion, or a series of single speakers.

Will your presentation be highly structured and formal? Do you want the focus to be on you at the front of the room? Will there be much audience involvement? Do people have to twist around in their seats if someone in the back row is talking? Will it be such a small room that some people may have to stand? Is the room so large that your audience will have difficulty hearing and seeing your material?

Sometimes conference centers and resorts send mixed messages with their plush facilities and task-oriented meeting spaces. The rooms in some companies are cold and impersonal with tile floors, bare tables, and hard chairs. In such environments, it will be difficult for the speaker to keep everyone's attention focused on the message.

One presenter told me that her association held a luncheon for 400 in a ballroom with music and an entertaining speaker. Immediately following the gala, she was responsible for presenting material to a group of 25 in a corner of the ballroom. It was extremely difficult for her and her listeners to make the transition and overcome the feeling that they were all lost in a huge, unfriendly space. Her best defense would have been to get the audience focused at once by asking questions, having them interact with each other, and involving them intellectually and emotionally with her material.

Tom Zimmerman, a well-known researcher in integrated circuitry from TRW, says that he arrives early at the speech site and walks around the room. He sits in the back row of chairs to get the same feeling his audience will have. He puts on a viewgraph or inserts a slide in the projector and walks back to the most distant row of seats to see if the participants will have a clear line of sight, and if the visual is legible.

If you will be speaking in an unfamiliar environment, request the dimensions and planned room arrangements with your host beforehand. *Be firm about your requests and submit them in writing.* Illustrate your desired seating arrangement and fax it back, along with typed instructions to the person in charge of arrangements. Make phone or e-mail contact for verification and find out who to contact on arrival.

Upon arrival, talk to the person in charge of the crew and give them your sketch of the room setup and instructions. Visit the room the night before, even if there is another event going on, to assess any problems. Be at the site early in the morning, as you may have to reset the room yourself.

There are several seating arrangements that are suitable for different types of meetings and presentations. The best configuration will depend on the number of attendees, the interaction you want from the audience, and how you feel most comfortable. These arrangements are listed below and depicted in the figure on the next page.

- *Curved seating.*

  A semi-circular seating design will improve audience dynamics, provide good visibility, and maximize the available space. Any time you have movable chairs, try a curved arrangement of seats. Eliminate the center aisle to capitalize on the energy between the audience and the speaker. Flare out aisles 45 degrees. Have breaks in the seating for every 12 chairs to permit easy access and reduce distractions from late arriving or exiting participants.

- *Office arrangement.*

  For a small setting in an office, place screen for best visibility and check for window glare. Be creative with available space, furniture, and electrical outlets. Take advantage of this intimate setting to encourage interactivity.

- *Classroom.*

  For small or large groups or longer sessions, work tables can be set up in front of chairs. Arrange the tables herringbone-style to encourage interaction between participants. Avoid having chairs directly behind one another.

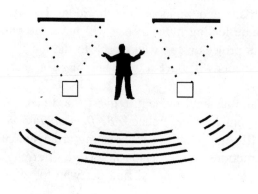

Curved

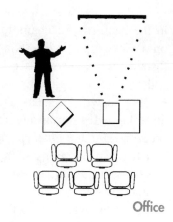

Office

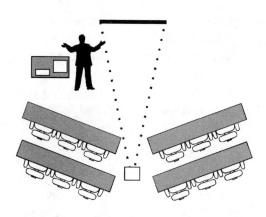

Classroom

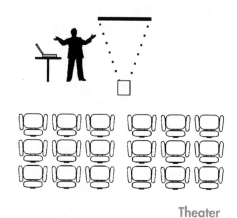

Theater

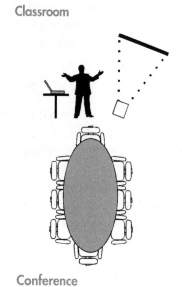

Conference

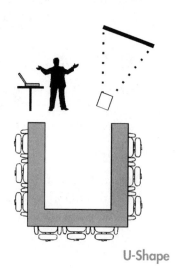

U-Shape

Courtesy of Shane Eckel

- *Theater style.*

This is sometimes referred to as auditorium style because the rows of chairs are placed much like those in a movie theater and may be fixed to the floor. A stage or platform is positioned in front, and the chairs may or may not be staggered. This configuration can be used for large groups, but seats lacking arm rests inhibit note taking. Be aware of the sightlines for visuals. You will probably need to use a microphone.

- *Conference room.*

When there is no primary leader, this configuration is useful for audience interaction and idea exchange. The speaker should stand to deliver the presentation. Allow at least 2 feet per person for elbow space, or more for comfort. The conference room arrangement is useful for four to twenty people, depending on the size of the room and table.

- *U-Shape.*

This arrangement is useful for up to twenty-four people. If you want your audience members to see and participate with each other, this is ideal. The speaker or facilitator can move around within the center and encourage involvement.

- *Banquet or rounds.*

This is a good arrangement for the exchange of ideas in small discussion groups. Position six to ten people around tables six to eight feet in diameter. This pattern will sometimes be uncomfortable for participants if they frequently need to return their attention to the speaker at the head table.

If you are to speak in a room that has been used for another event, change the seating arrangement so that it will be comfortable for your audience to listen. If the stage was set for the Shakespearean tragedy *Macbeth*, you wouldn't want to use it to perform the TV comedy *Frasier*. The proper setting can provide a background that will at least be neutral, instead of detracting from your message.

Take into account wheelchair access and up-front seating for the visually and hearing impaired. Realize that comfortable chairs may relax an audience too much, but if chairs are hard and stationary, the audience will have a tendency to fidget. Your listeners will be more alert if they are sitting in cushioned chairs that move.

If you have a choice of rooms, there may be tradeoffs. A moveable screen placed at an angle to the audience may be more important to your presentation than a room with windows. One presenter had an audience of 250 people in a very large room. Many people sat in the back, and he couldn't coax them forward. If you run into this situation, stack the chairs in the back or cordon off the area with a banner so that early arrivals are forced to sit closer. As additional people arrive, chairs will still be available.

Emotions travel quickly through an audience. If you feel that your meeting will be a *positive* one, try to get everyone *grouped together* to reinforce responses. If you have to announce *bad news*, it may be better to have the audience *scattered*.

ENHANCING THE SITUATION

The realities of the situation, of course, may leave you with little choice or flexibility regarding your meeting site. You may have to make do with whatever is available. If you are training at computer workstations, you are usually locked into straight rows. Try to include one or two exercises where the participants move away from their tables. Don't let inconveniences upset you. Concentrate on doing what you can to improve the situation and get your audience focused on you and your message.

• *Keep a clean house.*

Arrive neat and with as few supplies as possible. You can intimidate people by dumping stacks of folders on a front table or setting up a wall of high-tech equipment and tangled wires. Drape unsightly apparatus with table-cloths or position them away from your audience. Keep reference materials out of sight or people will spend their energy on estimating the time it will take for you to get through them. Dispose of used coffee cups, wastebaskets, or any nonessentials, and remove extra tables and chairs.

• *Protect yourself.*

One of my clients rehearsed for an interview on the 6 o'clock news. He told me later he couldn't read his notes because a sound man had crossed two microphones directly over his papers. "Couldn't you move them or ask someone to fix the mikes differently?" I asked. "Everything happened so quickly that I went along with what the crew told me to do," he answered. In such situations, it is *your responsibility* to request and insist on changes that will enhance your appearance.

You will find that some staff are reluctant to change arrangements in conference rooms or hotel meeting sites. One speaker asked to have a lectern moved to the side, but was told that was not possible because the microphone wires were already taped down. When she got down on her hands and knees and started pulling up the tape, the service staff decided to accommodate her request. If a call to management fails, it will be up to you to rearrange chairs, projectors, and other equipment.

• *Come out from behind the lectern.*

Podiums are objects you stand on; *lecterns* are objects you can place your notes on. I suggest you avoid lecterns, for they can create a barrier between you and the audience. However, if you are using notes in a conference room, a short tabletop lectern will allow you to have better eye contact with the audience.

• *Know your competition.*

Check on events scheduled at the same time in rooms nearby. I will never forget addressing a group of managers in a hotel room during the Christmas season. In the midst of my talk, a deep, resounding "Ho, Ho Ho!" came booming over the loudspeaker. I tried to continue, but we were interrupted with, "Little girl, what would you like for Christmas?" This was followed by whispers, squeals, giggles, and a recurring, "Ho, Ho, Ho!" I dispatched a class member to ask the management to fix the sound system, which alternately became louder, softer, then fell silent. I went back to my presentation, but five minutes later "Ho, Ho, Ho!" rang out again. I was able to get my audience to stop laughing and focused by asking them to do a two-minute communication exercise with a partner. I copied their comments onto a viewgraph and brought their attention back to the projected image in front of the room.

• *Thank you for not smoking.*

It is important to establish no smoking rules before a meeting starts. Most companies in the United States have no smoking policies, but do not expect this to be true for international meetings. There is nothing worse than having a presenter gasp and cough because of smoke or having to ask a key decision maker to snuff out a cigarette. One company placed this sign on the conference room door: *"If you smoke, we will assume you are on fire, and put you out."*

• *Keep the spotlight on you.*

Hotel lighting is notorious for making the presenter look like a visitor from Transylvania. If you check the lighting and find it less than flattering, move the lectern/table or stand in a more favorable spot. I recently attended a presentation on advertising on the Internet. The speaker had turned off the lighting in the front of the room so that we could see his on-line demonstration. He did a wonderful job of clarifying the complex material. Unfortunately, he never emerged from the darkness to establish rapport with his audience. The first time we really saw his face was when he came down the aisle to answer a question at the end of the presentation.

- *Group your slides or computer graphics so that the lights can be turned up between a series of visuals.*

  Know where the room's light switches are located and appoint someone to turn them on or off at your cue. If you are using handouts, make sure there is enough light to read them.

- *Work with windows.*

  Windows generally have a positive effect on meetings, but there are exceptions to the rule. If the scene outside is distracting, draw the curtains or blinds. One speaker told me she checked the meeting room the night before her presentation and discovered it was in the cafeteria. She neglected to note a wall of glass windows two stories high at the end of the room. The next day, the glare at noon made it impossible for the audience to see her visuals on the screen. She had to eliminate the visuals and change her whole approach, which she felt weakened her message. "I learned a valuable lesson," she ruefully acknowledged. "Now I either visit or visualize a facility at the *same time of day* as my presentation to appraise the effect of the lighting."

- *Check the acoustics.*

  The first requirement of a good speech is that it is comfortably heard by everyone. Have someone verify your voice carries to the back of the room or use a microphone. If there is the slightest chance of a problem, ask the audience if they can hear you *before* you begin. Audiences are reluctant to stop a speaker to ask her/him to speak louder.

- *Make your mike work for you.*

  Avoid using microphones if possible. However, use one if there is the slightest chance your voice will not carry adequately to the back of the room. If a small room is darkened for computer graphics or slides, a microphone will emphasize your presence in the room. Be selective in your choice of microphone. Wireless mikes are easiest to use because you need not worry about tripping over a cord and you are free to walk into the audience. Women might want to wear a jacket to conceal the transmitter. *Always test the microphone before you use it, because batteries frequently run down.* Since most wireless mikes use mixed frequencies, there is a chance they will pick up interference from taxis, rock bands, or radio signals. Be prepared with a backup mike.

  A lavaliere microphone goes around the speaker's neck on a cord which may need adjustment. Small clip-on mikes are also erroneously referred to as lavaliere mikes, so be exact when you request equipment. Avoid mikes that are affixed to the lectern. If you have no other option, position the mike the distance of your outstretched hand from your mouth. Hand-held mikes are trickier to use, since the volume of your voice can fluctuate as you move the mike. They require practice and can also present a problem if you have to move equipment, adjust viewgraphs with both hands, or use props or a remote control.

- Microphone cords should not be twisted or bent.

- Avoid touching the mike or tapping it in any way.

- If you have a cord, slip it through your belt and sweep the cord behind you.

- Arrive early and establish a volume level that can be heard by everyone in the room, but doesn't overwhelm the audience.

- Refrain from sudden bursts of breath or surprising changes in volume.

- Cover your mouth to the side if you must cough or clear your throat.

- Talk conversationally.

- Once you have the microphone in place and turned on, be careful about casual remarks unless you want them broadcast to the entire group.

- If you experience microphone feedback, the microphone is picking up the sound coming from the loudspeaker and may need to be moved or the volume adjusted.

In panel discussions, the participants are usually seated at a head table with one or more microphones on short stands. You don't have to lean awkwardly over the microphone because other panelists do. During an Oceanic Engineering Society/IEEE Association meeting, one presenter took the microphone out of the stand, got up, and placed a viewgraph on the projector. He discussed it briefly, then turned the projector off and continued to stand for the remainder of his talk. It was a smart tactic since his body language, voice, and energy were much stronger when he was standing.

- *Respect time limits.*

Position a clock with large numerals within your range of vision and adhere to your allotted time. Some speakers wear their watch facing inward so they can unobtrusively check the time. It is pointless to go beyond your time limit. You will only irritate the listeners and other presenters. Even if members of your audience have their hands raised for questions, conclude the session on time and offer to answer them afterward.

• *Control the temperature.*

Try to have the room temperature between 68 and 70 degrees F. If it is any warmer, the audience will have a difficult time staying alert. Fluctuating room temperatures disturb the audience and physical discomfort will be foremost in their minds. Either find out how to adjust the temperature yourself or ask who can do it for you. Do not wait until you begin to have problems.

An executive, a client of mine, was scheduled to give a one-hour speech on the second day of a global conference in New York City. The air conditioning was not working and the room began to get very warm. By 2 p.m., the audience of five hundred had begun to drift in and out of the ballroom. Others were sleeping or engaged in small group conversations. Noise drifted in from the hall. Some of the international speakers spoke slowly in heavily accented English. When controversial questions were raised, several speakers suggested that my client would answer them.

By 3:45 p.m., the room was unbearably hot, and the sweating audience was anticipating a break at 4:00. I walked up to the executive and said, "Capsulize your speech into fifteen minutes. Give your introduction and the first anecdote, hit the three main points, and tell everyone you will meet them in the bar to answer their questions. The audience has a press release of your speech and can read it in its entirety if they want to."

When my client was introduced, he began, "I'm going to make my remarks brief and have you all out of here in fifteen minutes." The audience became alert and attentive and, when he finished and announced he would meet them in the bar, they cheered. There, he fielded questions brilliantly for an hour and was the hit of the conference.

• *Take charge in emergencies.*

I was finishing my closing remarks during a breakfast meeting when a man suddenly stood up in front of me, gasped for breath, grabbed his throat, and fell to the floor. Everyone was so startled that no one reacted for a few seconds. I was waiting for someone to get help when I realized that I was the one in charge at that moment. Using my microphone, I asked if there were any doctors or nurses in the room, directed a person by the door to call 911, announced that the meeting was concluded, and asked everyone to leave at once. The medics arrived within minutes and revived the man with oxygen. It was a frightening situation.

As presenter, you may find yourself faced with an emergency and discover that it is up to you to take charge. Always locate emergency exits and follow them to the outside to determine if they really are clear passageways. Know where and how aid can be obtained.

SPEECH SITE CHECKLIST

1. Seating Arrangement
   - ____ Curved row seating
   - ____ U-shaped classroom
   - ____ Office meeting room
   - ____ Theater style
   - ____ Conference style
   - ____ Classroom style

2. Audiovisual Equipment
   - ____ Flip chart, colored pens
   - ____ Whiteboard or dry erase board
   - ____ Overhead projector (spare lamp)
   - ____ Table for projector/viewgraphs
   - ____ Computer
   - ____ Extension cords
   - ____ TV monitor(s) for computer or video
   - ____ Microphone (stationary, wireless, hand-held)
   - ____ LCD panel or projector
   - ____ 35 mm slide projector
   - ____ Screen
   - ____ Video camera
   - ____ VCR
   - ____ Remote control

3. Room Setup
   Who will deliver the audiovisual equipment _____ Phone # _____
   Time I can check out room _____ Time of audiovisual setup _____
   Placement of above items, i.e., screen to the left of speaker, etc. _____
   Lectern _____ Placement _____
   Location of electrical outlets, light switches, extension cords _____
   Adequate lighting on speaker _____ Who will dim lights _____
   Room temperature 68°F ____
   Emergency exits _____ Restroom facilities _____

4. Handouts _____ Number needed _____ Placement _____

5. Supplies
   - _____ Clock or watch
   - _____ Masking tape
   - _____ Pencils, pens, markers
   - _____ Note paper
   - _____ Name tags
   - _____ Water for speaker

- *Create the mood.*

Even before the presentation begins, a speaker can divert attention from site problems and warm up a room by exhibiting interpersonal communication skills. Greet and talk to audience members as they arrive. It is a mistake to be fiddling with papers or audiovisual equipment. Introduce yourself and shake hands. Make introductions between other audience members. Offer small talk or discuss the purpose of the meeting and ask what information would be helpful to them. Then excuse yourself to make any final physical or mental preparations before you begin.

If you have any control over when the presentation begins, start on time even if there are only a few people present. Of course, if the key decision maker has not arrived, you will need to delay your speech.

PLAN A, PLAN B, PLAN C

Murphy's Law will, at some time, go into effect before or during one of your presentations. Call ahead and talk with the technician who will be setting up your equipment. Make sure he or she will be available before and during your presentation. A friend of mine contacted the audiovisual person and was assured her special equipment would be available for her 9 a.m. presentation two months later. Unfortunately, the technician failed to mention he would be on vacation that week. The hotel staff knew nothing about her special requests. Anticipating problems and having a backup plan will give you more control, but not every contingency can be predicted.

Your level of preparation and comfort with the site arrangements will have a critical effect on how you feel about yourself and, consequently, your delivery. Do what you can to improve or maintain the temperament and attention of the audience, but realize that you may need to change your style of speaking in response to the environment.

KEY IDEAS

- Take responsibility for your presentation site; don't depend on anyone else to "set the stage" for you.

- Triple-check all equipment, supplies, and room arrangements. Have backup plans in mind, especially for microphones.

- Arrive early so you can change seating arrangements if necessary, or reposition the audiovisual equipment.

- Minimize distractions of noise, sunlight, intercoms, and temperature.

- Appoint someone to troubleshoot during your presentation.

Chapter 19

CREATING FAVORABLE INTRODUCTIONS

"It gives me great pleasure...."
—Anonymous

OVERVIEW

This chapter gives you tips for your introduction, which will help establish your credentials and whet the appetite of the audience. Your introduction should include details about your background pertaining to your topic and audience. It should substantiate your expertise and trustworthiness, and set in motion the audience's acceptance of your ideas.

Television stars know the value of a warm-up comedian before their appearance; you can also benefit from a bright and, perhaps, witty lead-in to your speech.

Orson Welles was lecturing in a small Midwestern town before a very sparse audience. He opened his remarks with a brief sketch of his career, saying, "I'm a director of plays and also a producer of plays. I am an actor of the stage and motion pictures. I write and produce motion pictures and I write, direct, and act on the radio. I'm a magician and a painter. I've published books; I play the violin and the piano." At this point, he paused and, surveying his audience, remarked, "Isn't it a pity there's so many of me and so few of you?"

One presenter told me, "I used to be somewhat cavalier about introductions, but now I write my own. I have had dreadful experiences with moderators who have ruined my credibility, and I have had to spend the entire speech trying to get it back." And a University President confided, "Indeed, on some occasions after introductory remarks that seemed more like eulogies than introductions, there have been times when I have felt that the only truly appropriate thing to do is to die." *I strongly advise everyone to write his or her own introduction.*

Send your introduction ahead if it is a formal occasion, but always carry one with you for extra insurance. You will have more control over the information presented. If you send a biography, expect to hear it read word for word. You still need to send an introduction to properly set up a speech for a specific audience.

Arrive early at the speech site and immediately ask to meet the person who will introduce you. Give the person your written introduction and review the proper pronunciation of your name and the title of your speech. If you have researched your audience and feel it is important to include your specific background in artificial intelligence or genetics, ask the introducer to precisely follow your written material. If appropriate, you could suggest the introducer briefly relate how she or someone in the organization was a former colleague or how she came to know about your work. If you have an accent, you might want your country of origin mentioned in the introduction so the audience isn't distracted by trying to guess where you are from.

Some speakers with celebrity status may prefer to remain hidden and make an entrance when introduced. However, the audience is able to relate to you and your credentials if you are visible when your introduction is being read. Watch how the audience responds, as it may yield important clues regarding how you should proceed in your opening remarks.

Within a company, there will usually be no formal beginning of a business meeting. Someone will simply say, "John, let's hear from you," and you are in the spotlight. If everyone knows you, there is no need for you to give your background, but there is a need to have the audience focus on your subject.

Some scientific and technical audiences will not pay full attention unless they are acquainted with your credentials. A moderator may insist on delivering the introduction he or she wrote. If it says little about your experience, it is perfectly appropriate for you to begin, "Let me tell you about the background (or expertise) that I bring to this subject." Be brief and matter-of-fact. If the moderator makes a grievous error or omission, you can easily slip into your remarks, "I wish I *had* been part of the Board to establish those regulations but *my role* was only as a observer..." or "During my four years as a programmer at Netscape...." It is a good practice to include a brief biographical sketch and picture at the end of your handouts. You can make reference to this, especially if the introduction has not established your credibility with the subject. This is just as valuable if you are presenting at an in-house face-to-face meeting or company videoconference. Your peers may be unaware of your achievements.

Why are *you* speaking at *this particular time* on *this particular topic* to this particular audience? When your background is described and you are introduced as an expert, you need only to establish your authority in relation to a certain aspect of your subject. It is not necessary to document your life after the high school debate team, but it may be important to mention a significant project or paper.

Include a personal note in your introduction, such as, "she enjoys mountain climbing" or "he recently wrote an article for *The Industrial Physicist* magazine." This will help to establish rapport with your listeners. A witty line will have your audience smiling as you step to the lectern. Your introduction should place you in an advantageous position for starting your presentation.

KEY IDEAS

- Stimulate your audience's desire to listen.

- Establish credentials acceptable to the specific audience.

- Don't protest too loudly or react too humbly at a flattering introduction.

- Review your introduction with the moderator.

- Write your own introduction. Keep it crisp and short.

Chapter 20

VIDEOCONFERENCING

"There is so much information being beamed around this world that it's almost as if there is another layer of the atmosphere. And it is going to become easier and easier to breathe that air no matter where you live."
—Irving Goldstein

OVERVIEW

A videoconference is a fast, efficient delivery system that allows companies to disseminate information in real time to employees, sales representatives, customers, and the general public. In this chapter you will learn techniques to help you become a compelling presenter or facilitator of your videoconference. You can be calm, comfortable, in control, and enjoy communicating through this demanding but exciting medium.

If you haven't already been a participant in a videoconference, it is a certainty that you *will be* in the near future. You may be interviewed for your next position or evaluated for promotion through videoconferencing. Some of your training will most likely be delivered via distance learning. You may be part of a room-sized gathering, or seated at your desktop conversing one-on-one with people from around the world. Success will depend upon your developing new attitudes and learning new skills to exploit this exciting and increasingly popular technology.

Videoconferencing has come of age and is now accepted as a mainstream communication tool. The term "videoconferencing" can apply to many different combinations of audio and video communication.

- *One-way videoconferencing*—viewers at multiple sites can watch speakers at the originating site and interact by telephone.

- *Two-way videoconferencing*—allows participants at the originating site and receiver sites (or downlinks) to both see and speak to one another.

- *Desktop videoconferencing*—runs on your PC and allows one or more individuals to collaborate, and can also connect to room videoconferencing systems. You and your colleagues can share a common workspace on the screen; retrieve text, images, and video clips from remote databases; create material; simultaneously edit and annotate pictures, charts, spreadsheets, word processing documents, or scanned images; and merge it with the rest. When you are finished, each individual can send the report or presentation material with voice annotations and visuals to others on electronic mail, or post it on their Web site.

BENEFITS OF VIDEOCONFERENCING

Why is videoconferencing becoming so commonplace? Advances in technology have lowered costs and made available broadcast-quality video with realistic sound. Videoconferencing can be used to build an accurate, relevant, and timely communication infrastructure. Organizations are becoming increasingly complex, global, and decentralized, and this medium can deliver a message to targeted audiences without considering the boundaries of geography. Institute for the Future Director, Mary O'Hara Devereaux, says, "You can do a lot more from a distance, and the sheer volume of international work is increasing. Now, supported by technology, work can happen twenty-four hours a day because the wider dispersion of global workers means a team of workers can span time zones."

Companies such as Federal Express, General Motors, and Hewlett-Packard have employed videoconferences to train and retrain employees, introduce new products, communicate with suppliers, promote special events, and inform employees worldwide about changes in top management, supervisory procedures, and policies.

Boeing trained 21,000 people over a six-month period using satellite broadcast services linking sixty-two learning sites nationwide. Boeing producer/director Darrell Prowse commented, "We successfully created an intimate classroom feeling with trained on-site facilitators who led small groups through interactive exercises. Key executives, presenting both live and on-tape, were seamlessly integrated into each of the two-day conferences. It was a very structured program." This interactive two-day workshop also will be delivered to an international audience of Boeing managers in Great Britain, Germany, Japan, Australia, and the People's Republic of China.

Engineering, scientific, and high-tech firms find videoconferencing efficient for creative brainstorming and problem solving. A facilitator has the opportunity to include far-flung participants having a wide variety of perspectives and expertise. He or she can also reconvene members of a group so that one video-conference starts an incubation process for innovative contributions to the next session. In contrast, when you travel to another city for a day of meetings, many problems are inadequately resolved because you need to adjourn the meeting and catch a ride to the airport.

Manufacturing companies can use the medium for problem solving when designing or testing parts. Engineers can show and annotate blueprints and detailed engineering diagrams, as well as display three-dimensional models from site to site. This medium has helped engineers and managers use time more effectively, shorten developmental cycles, accelerate decision making, and move products to market more quickly. It is extremely useful in cementing strong customer relationships, because it delivers "just-in-time" information, improving responsiveness to customer needs, whether they are across town or across the world.

No one is excited about the prospect of attending additional meetings. But videoconferencing can give an organization exceptional returns in productivity while saving time, money, and energy. Many companies speak of the exciting synergy generated by electronic meetings. *Videoconferencing can be more effective than face-to-face meetings*—but only by *design*; not *chance*.

Your Videoconference Should Be VIVID

Your participation and responsibilities in a videoconference will be dependent upon the particular level of videoconferencing, which might be:

- *Desktop videoconferencing—you* are totally in control of time, camera, visuals, props, etc.

- *Room videoconferencing—a room operator* may control and preset the camera, run the document camera, and carry out other duties. Or *you* may have responsibilities surrounding your presentation.

- *Event videoconferencing*—may be on location, outside, or in a ballroom, but entails a *full production staff.* You will preplan and, usually, work from a script.

Video requires close attention to delivery skills. At first, the unfamiliar process of a videoconference can be intimidating. Any number of important people may be watching. There will be no retakes. Dr. Philip Westfall, director of the Air Force's Center for Distance Education, explains, "The satellite broad-cast tends to sharpen an instructor's overall presentation, as well as their delivery. An unprepared program is magnified on the air. It forces the teacher to more closely analyze their instruction material because once that on-air light goes off, there's no time left to wrap up the day's lesson."

The acronym **VIVID** describes the elements you need to remember for your compelling videoconferencing presentation:

- **Viewers.** Design your material from your audience's viewpoint. Target your *objective*, tailor your *message,* and include the *profit-value* for the audience. Do a thorough *audience analysis* of the viewers with your checklist (see Chapter 6). Language barriers may demand that someone who is multilingual be available to troubleshoot. Time zone differences can be challenges for scheduling and keeping attention, especially for international conferences. Even in North America, a lunch break on the east coast would be a stretch break on the west coast.

- **Interactive.** Design and deliver your message so that your audience becomes *involved mentally*, *physically,* and *emotionally*. Make the audience aware of your expectations for interaction. Participants can be reluctant to join in a discussion after your presentation. Instead of asking, "Are there any questions?" preplan a question from an individual at a specific site to start the discussion. Suggest what to look for in a video or visual, and then say you will be asking for feedback. Encourage use of the telephone, keypads, whiteboards, and faxes. Be creative!

- **Visuals.** Think in *images*. Visuals reinforce and clarify your points, aid viewer retention, and add variety to your presentation. As you do research, think how the information can be translated into images, using graphs, cartoons, photos, or preproduced video. Bigger is better! Pay special attention to the *legibility* of visuals. Handouts play an important part in a videoconference and should reinforce your message. This is an ideal medium for showing models and props as you can zoom in for close-ups with a document camera.

- **Immediacy.** One of the biggest advantages of videoconferencing is that the speaker can relay up-to-date information to viewers. You will need flexibility to incorporate this data, visuals, or props while maintaining rapport with your audience and preserving the cohesiveness of your speech. You may have to fax last minute handouts, or improvise by inserting a whole new sheet of financial figures you've just downloaded from the Internet.

- **Delivery.** Dynamic delivery skills are a must, because video can make you appear listless. Clearly define your image. An expressive voice and energetic body language will attract the audience's attention and encourage interaction.

THE CAMERA AND YOU

Video can be a demanding medium. Audiences expect polished delivery skills, and they will watch to see how you handle controversial questions and how you respond to comments. Your listeners typically will decide in the first few minutes whether they will give you their attention, and that decision will be based on the value of your material, your delivery skills, and whether they feel comfortable with you. Your listeners are accustomed to switching channels. There must be something to interest or challenge them.

- *Speak in a conversational tone, but with authority and conviction.* Video is an intimate medium. You will be most successful if you imagine your audience as a few friends sitting just a few feet away. The camera is *your friend.* It is your messenger to your listeners.

- *Eliminate meaningless gestures.* In a face-to-face meeting, the visual field is diffused and the surroundings provide distractions for the audience. But when the audience is focused on a screen and a close-up of you in a box, meaningless gestures and negative mannerisms will be exaggerated and can distort, dilute, or confuse your message.

- *Call up images and ideas in the minds of your listeners at other sites—* without affecting an artificial eagerness—and you should succeed. Enthusiasm in your voice and body language comes from a desire to share what you are saying, and a belief that the information is important to your listeners. Your energy should come from your solar plexus, not just from the head and neck. *Avoid being a talking head.* It is especially important during international meetings to speak slowly, using precise articulation and meaningful gestures.

- *Know which camera is focused on you and whether it is a long-shot or a close-up.* Review hand, verbal, or sound cues with your production staff. Plan when you will cut to a slide, roll-in, or document camera.

- *Keep track of time.* Strictly adhere to the schedule. Going three or four minutes overtime may eliminate another speaker's remarks. Allow time to give a summary, assign follow-up activities, or schedule the next videoconference.

- *Act as if the camera is on you*, once the conference has begun. Don't lapse into reverie, stare at the monitor, or allow your attention to wander. Be an active listener.

- *Use good posture; sit up straight.* Do some isometric exercises or deep breathing to give your voice and posture a boost of energy.

- *Own your space.* The video camera absorbs energy like a black hole gobbling up stars. Have your presence expand until it takes up the entire room. The key is radiating positive energy and enthusiasm even when you don't feel it. Maintain a *relaxed alertness,* as your listeners will mirror your body language. If you lack energy and enthusiasm, you will actually tire your audience. Project the feeling that you would rather be attending this meeting than doing anything else in the world. Just before the start of the videoconference, visualize yourself at a time when you felt calm and in control. Remember an occasion when you felt good about yourself, and draw on that energy. The camera can highlight your naturalness and ease.

Jeff Raikes, Senior Vice President of North American Sales at Microsoft, is an example of someone who "owns his space." He hosts videoconferences at Microsoft for sales reps with a strong, authoritative presence, yet skillfully encourages interaction and questions from his viewers. He is particularly adept at facilitating a relaxed, conversational exchange of information from a panel of speakers, but still keeps a brisk pace to the proceedings. Raikes says he approaches his videoconferences with a sense of the key issues that need to be communicated to his audience. Then he outlines the questions for his panel that will bring out those issues.

THE CAMERA NEVER BLINKS

This may be the only time that clients or colleagues at other sites see you; they will make an immediate judgment. Project an image of professional competence. The camera can add ten pounds, which should provide incentive to start those stomach crunches and skip the junk food. Choose clothing that is appropriate to your industry, but wear your "casual best" if there is no dress code. The situation is much more relaxed if you are communicating by desktop; however, a sloppy appearance will detract from your credibility. Large hair styles can draw too much attention. Give yourself a manicure if you are going to be handling objects under a document camera.

Nearly everyone struggles to meld the image they hold of themselves in their heads and their actual appearance on a video monitor. It takes time to accept how we appear to others. Video does not give us a mirror image. Sometimes people will get so distracted watching themselves on the monitor and lamenting their thinning hair, wrinkles, or their attire that they lose track of their thoughts. Try to critique your physical appearance objectively and focus on how well you communicate your ideas.

Watch commentators and TV personalities. Study their styles and note specifically what it is you like and dislike about them. Learn from them. However, be wary of imitating your favorite newscaster. What can you incorporate into your own style? Practice "as if" you were a lawyer speaking to a jury, or a confident, enthusiastic professor instructing a class. This exercise can bring out qualities that you may hide. Develop your own unique, special style. Remember, *YOU are salable*!

The most important factor affecting how you communicate is *how you feel about yourself*. Like yourself! Believe in yourself, your ideas, and the value they have for other people. Your listeners will hear and see this. Have fun and enjoy being on camera! Smile—let your personality come through.

Do not get caught up in being perfect. Video is a very powerful but difficult medium. A creative, informative presentation won't always be "cosmetically perfect." Visualize yourself successfully communicating on-camera. Imagine it from beginning to end, and live it with all your senses.

The camera can be your ally. It can intensify your personality. It can reinforce the positive elements of your presence and delivery. It can showcase naturalness and enthusiasm. After a few videoconferences, participants agree that the technology becomes transparent. They feel comfortable participating in electronic meetings and begin to focus on finding creative ways to increase productivity.

FACILITATORS

My own survey of videoconferencing users shows that the most important ingredient to a successful conference is a strong facilitator, someone who has the interpersonal skills to focus and direct a meeting toward its goal. What if you are asked to facilitate the electronic meeting? You should solicit agenda items in advance, distributing the final agenda prior to the meeting. This agenda, by clearly defining a framework, will cut needless talk and keep the meeting moving toward its destination. The meeting facilitator is well advised to plan major contributions and to assign responsibility for their execution to key personnel.

The leadership style that worked in last week's face-to-face meeting will not always work in a videoconference. The reason: Replacing face-to-camera interactions changes the meeting's dynamics. You will need to establish a good rapport with key people at other sites so that the meeting runs smoothly and brings out the best in all the participants. Your leadership style will set the "tone" of the meeting. If it is a multi-point meeting, it will be more of a challenge to run it smoothly so that everyone is included. Keep the meeting moving along toward its goals within the time limitations.

Some facilitators will only have the responsibility of running the meeting, but other situations will require someone adept at multi-tasking. You may need to know how to operate the camera for different viewpoints, be able to use a document camera, do presets, push mute buttons, or handle phone calls. A facilitator should know some basic troubleshooting if the audio or video fails, or how to get technical assistance. If it is a highly produced event, you may need to practice wearing a separate communication line (which fits in the ear and is called an IFB) for receiving instructions from the director.

If a strong facilitator takes the reins and all participants come prepared to contribute to the agenda, everyone can emerge from a videoconference feeling they have accomplished something, their time was well spent, there is a sense of team spirit, and they have a clear direction of their next step in a project. How many of your recent meetings have generated that same sense of purpose and satisfaction?

TIPS FOR FACILITATORS

1. Start on time.

2. State purpose of videoconference.

3. Explain how and when interaction will take place.

4. Introduce participants.

5. Review agenda.

6. Encourage participation and moderate questions and answers.

7. Stimulate critical evaluation.

8. Raise issues. Guide group thinking.

9. Maintain focus.

10. Clarify attitudes and restate them.

11. Resolve conflicts.

12. Summarize meeting's progress.

13. Conclude with clearly defined action steps and time frame.

14. Schedule next videoconference.

15. End on time.

REHEARSE YOUR PERFORMANCE

Since they rarely prepare for face-to-face meetings, some people feel they don't need to rehearse for a videoconference. Remember our previous discussion about "home court advantage?" Rehearsal and familiarity with the equipment dramatically increase your chances of a successful videoconference. Video broadcasts are usually tightly scheduled. Memorize your opening lines and the beginning and ending of each segment. You will need to get to the point and finish within the allotted time. Although it should look spontaneous, the event or the meeting must have a definite structure, move along at a comfortable but brisk pace, and come to a conclusive end. Thorough preparation will help coordinate the roles of the participants and minimize wasted time and inefficiency.

Request feedback during run-throughs or videotape your presentation. Some people are unaware they frown, squint their eyes, appear mad, or smile all the time. Others use vocal fillers such as "um," "you know," and "ah." Correct the mannerisms that may annoy your audience and intrude on the quality and effectiveness of your delivery. If TelePrompTers are available for special event videoconferences, rehearse with them. Practice incorporating your props and visual aids into your presentation.

It is critical to review visuals with the video director. Take a dry run using a laptop computer or any other equipment. Note where you will be referring to handouts that your listeners will be using at the receiver sites. Visit the room where you will be facilitating, presenting, or participating. Check out preset camera angles.

If it is impossible to rehearse at the site, write out your objectives and what you believe to be the expectations of the audience. Finalize an agenda. Learn the proper use of the equipment. Have realistic expectations of what can be accomplished with this technology and capitalize on its real strengths.

DESKTOP VIDEOCONFERENCING

Advances in technology have made videoconferencing more convenient, portable, and cost effective. Plug-and-play desktop videoconferencing has entered a new era, with data and broadcast quality video transmitted over telephone wires.

President Roosevelt's fireside chats were held in comfortable settings. Your desktop videoconferencing actually has several similar elements. You can be seated in a relaxed manner in your chair, presenting to one or more listeners simultaneously while they view your slides, your whiteboard, or your latest video. You don't have to worry about your physical stance or what to do with your hands. The camera can be positioned in such a manner that your listeners aren't aware that you are glancing at your notes. You can even surround your-self with support staff. For example, during an important proposal, one firm had their marketing specialist at a whiteboard in the sightline of the presenter. The specialist could hear and see the presentation and contribute additional statistics, write questions, and bring up relevant relationships. Having resources at your fingertips can reduce the pressure, and you can concentrate on your key issues.

If you have a split screen with graphics and are not able to see all of your listeners, it is harder to adjust to their feedback. You can acquire a false sense of security, because you may think they understand. This scenario requires an in-depth audience analysis so you can anticipate your listeners' familiarity with the material and potential reactions. It is harder to control the focus of your audience with desktop videoconferencing. Consider what will be filling up the rest of the screen when you are speaking in the corner of the computer monitor. If you only have three visuals, how long will they remain on the screen? Can you refer participants to Web sites or other information? Find out if your visuals are positioned to your left or right at other sites so you aren't pointing in the wrong direction or have your eyes staring off into space.

Sitting at your desk may be relaxing, but you can also fall prey to letting your energy completely dissipate. Watch any tendency to slump, as it will affect the confidence and authority in your voice. Desktop videoconferencing can be so casual that participants aren't prepared or have a tendency to ramble. *Stay focused, deal with issues,* and *come to conclusions* in a timely fashion. Many of the previous tips on delivery skills for room videoconferencing are also appli-cable for desktop videoconferencing.

NEW SCENARIOS FOR VIDEOCONFERENCING

The recent design of videoconferencing rooms that are mirror images of rooms at other sites gives the feeling that you are sitting eye-to-eye with someone in the same room. Participants sit at a conference table that seems to continue into the visual image of the other conference table and *life-size images* of the attendees. The technology becomes transparent and it seems as if everyone is meeting in the same room.

You may find yourself being projected on a *thirty-foot screen* which will require even more attention to your delivery skills. Emerging technologies give movie theaters the capability to simultaneously transmit live satellite broadcasts to audiences in cities across the country or the world. With the use of a two-way camera at key locations, as well as electronic polling, the conference experience is made more participatory, thus enjoyable and memorable. A company can use its top people and everyone sees the same real-time presentation. Microsoft, Pepsi-Cola, Volvo, Autodesk, AT&T, and General Motors have all held a variety of meetings—from product launches to training seminars and corporate events—in this dynamic new setting.

Videoconferencing can become one of the most efficient and cost-effective ways of doing business. But electronic meetings shouldn't try to replicate face-to-face meetings. After all, if you candidly appraised the last meeting you attended, you probably wouldn't want to use it as a model. It is important that all participants understand the purpose of the videoconference, and how this particular event fits into the larger picture or long-range goals. A videoconference may go well, but if there is no follow-up or action taken, it was just a conceptual meeting and a chance to see Charlie in Atlanta.

It is now possible to interact with almost anyone on earth with a live video image by means of the Internet or a series of three communication satellite transmissions. Videoconferencing provides many solutions in the cycle-shortening and cost-reduction demands of our business world. Face-to-face meetings may become a luxury that we can't afford. Become proficient using this exciting medium. It can extend your influence as an engineer, scientist, technologist, or business professional.

KEY IDEAS

• Decide to excel as a facilitator/presenter for electronic meetings.

• Use this medium to its advantage and think in images.

• Be a prepared participant and make concise, up-to-date contributions.

• Talk naturally but maintain high energy.

• Rehearse. Become comfortable with your appearance on camera.

Chapter 21

COMMUNICATING IN A
BORDERLESS WORLD

*"Be ready or be lost. If you don't think globally,
you deserve to be unemployed and you will be."*
—Peter Drucker

OVERVIEW

*Speaking before an international audience can be a challenge for communica-
tors of technical information. The inherent difficulty of getting a message across
is compounded by linguistic and cultural issues. A thorough audience analysis,
extra care in the design and display of visuals, sensitivity in using humor and
personal anecdotes, and a clear structure will bring enlightened looks on the
attendees' faces. You will also avoid embarrassing faux pas. Become familiar
with social customs to guide your physical behavior. They will also help you
interpret the audience's response to your content and delivery. Though you may
view yourself solely as a company representative, multicultural audience members will also think of you as your country's ambassador.*

You have been asked to represent your company at an international conference
in Paris. The conference will have attendees of many nationalities and cultural
backgrounds. How will you reach your objectives or be persuasive in a
multicultural environment? How will you tackle the challenge of representing
your company and your country? Let's review how you would apply techniques
from this book in light of the different challenges presented by an international
audience.

A World of Similarities, A World of Differences

Many countries who have shared a cultural past with the U.S. (United Kingdom, Germany, France) have evolved in relatively similar directions. Save for the occasional language barriers, the interaction between speakers and audiences of such countries are nearly indistinguishable from that between American speakers and audiences.

A country that is *economically* westernized may not be *culturally* westernized as well. Indeed, U.S. cultural imperialism may have homogenized external factors such as dress, and soft drink preferences, but has not gone so far as to modify mores and customs. A Bedouin's oil wealth may buy all the trappings of western success, yet underneath it all, he may be as traditional as his father.

Although you may be alert to different countries with different sets of rules, remember that differences also exist *within* each country. Just as you would need to take into account different audience reactions in South Dakota as compared to Washington, D.C., be prepared for differences between Moscow and Trevier.

A global audience will expect a representative who has status within the organization. Otherwise, your message may be dismissed as unimportant. Your choice of spokesperson will convey the extent of your interest, or lack thereof, when you are dealing with an individual, a company, or government.

Attitudes: U.S. Informality vs. Foreign Formality

Speakers from other nationalities do not always take a democratic approach to giving a presentation. They assume a superiority because they are the ones dispensing the information. When the audience and speaker clearly know the "pecking order," and behave accordingly, all can relax. Americans, on the other hand, only relax when the parties involved are perceived as equal. They do everything to minimize the "Frog and Prince" syndrome, whereas other cultures are quite accepting. Rather than being too impressed, Americans are apt to believe that they are every bit as good as the speaker. But foreigners judge the immediate presentation, which can foster inequality. Meeting that person in a different situation, the hierarchy may change between foreigners, but remains fixed between Americans.

In some Asian cultures, people may agree with you for the sake of politeness, while having no intention of conducting business with you. The opposite is true of other cultures, where the audience is afraid to show acceptance of the message despite being completely sold, for fear of giving an impression that may be later contradicted by a superior.

A foreign audience may perceive humor in a technical presentation as unprofessional and the speaker as frivolous or boastful. Sharing a good laugh creates too much familiarity, therefore discomfort. Whereas we encourage a sense of humor which relieves tension, it can create tension abroad. Drawing on a personal experience to illustrate a concept can also be frowned upon. Disregard my encouragement to do this if addressing a global audience.

In the presence of an audience, colleagues who would in private address one another by their first names, revert to the use of Mr. or Ms. So-and-So or their title. Foreign speakers sometimes spend much time praising their colleagues (particularly their superiors), the company, and its accomplishments as a means of establishing credibility vis-à-vis an audience. Many times an employee's identity is tied to a company's identity, especially in Asian companies. Find out who you should recognize and their titles before you begin your presentation.

AUDIENCE ANALYSIS

While conducting a thorough audience analysis, it is important to:

- Accurately determine English language proficiency of members of the audience.

- Uncover cultural icons or taboos.

SIMPLICITE, SIMPLICIDAD, SIMPLICITY

Most presentations worldwide will be done in English. Unfortunately, this lulls the presenter into a false sense of comfort and confidence that they have the rapt attention of the audience. It takes an enormous amount of effort and energy for your listeners to attend to a brief technical presentation in a different language. The difficulty is compounded if they have had to sit through several days of a conference. Most countries have been exposed to MTV and Hollywood action movies. They have just as short attention spans as we do. They want to hear your message and move on!

- *State your premise* or *findings/conclusions* in the beginning of an oral presentation.

- *Preview your content.* This will aid listeners with English as a second language. The old adage of "tell them what you are going to tell them, tell them, and tell them what you've told them" is particularly effective with global audiences. If the listeners don't understand your point on the first go-around, you are ensuring they can assimilate it later. Repeat but don't rephrase. If you start using one term, do not substitute a synonym further along in your speech, as it will confuse your listeners.

- *Use unequivocally international references* such as a Nobel Prize winner, an internationally published author, or a well-known historical figure. You will be impressed by many audiences' familiarity with American history and personalities, but don't assume this is true for everyone. Briefly explain position and expertise of sources of lesser-known testimonials.

- *Persuade with logical appeals; use emotional appeals sparingly.* If you make a statement, back it up and know where to find supporting documentation.

- *Define and give the words of acronyms; seek out the local accepted acronym,* i.e., OPEC (Oil Producing and Exporting Countries) becomes OPEP in French (Organisation des Pays Exportateurs de Petrole).

- *Avoid idioms or nonliteral expressions.* During an introduction, one moderator mentioned that the keynote speaker "wore many different hats." A confused member of the audience assumed she must have an extensive wardrobe of accessories.

- *Use an unmodified verb instead of a verb followed by an adverb or adjective. Looking over* your calculations can be confusing, whereas *reviewing* your calculations will be understood. *Looking up* information on the Internet is not as specific as *searching* the Internet.

- *Use a noun or proper name instead of the pronoun.* Say "the software," "the electrical experiment," instead of *it.*

- *Select analogies and metaphors that will be familiar to the audience* or you can *decrease* instead of *increase* understanding.

Visuals

Visuals are essential and can help clarify complex information as well as aid retention. Images and graphs provide a welcome respite from a barrage of English.

- *Establish rapport and let the audience see you* before darkening a room for slides, multimedia, or videos. Keep the lights on as much as possible so that your audience can see your facial expressions and gestures, and you can respond to puzzled looks and stop to restate your message.

- *Choose visuals that clarify meaning, but are not offensive.* Imagery and clip-art icons may convey shared meaning in the U.S., but can sometimes transmit unintended messages through color associations, symbols, and quick edits to a foreign audience. You run the risk not only of conveying the wrong meaning, but no meaning. For example, in the U.S., a lightbulb symbolizes electricity, or on a more abstract level, an idea; to another culture, it may be just a lightbulb. Eliminate nonessential clip-art images, logos, grid lines, and textured backgrounds, which distract from the main message of your visual.

- *Carry your equipment if possible.* Keep in mind issues such as electrical current (carry an adapter) or lack of power. *Have low-tech visuals on hand.*

> • *Use maps when discussing distances,* as your audience may not know
> geographical locations.

Handouts

> • *Distribute an agenda or summary.* You will win extra points if you have
> a version in the native language of your host, as well as English. Some
> countries will expect their handouts tailored to the specific discipline of
> the recipient, which will require multiple versions of your handout. Unlike
> Americans, most foreigners still "worship" the written word, and will read
> every detail of your material. This can be advantageous for reaching your
> objective, because a nonnative speaker's reading proficiency is generally
> superior to his or her listening comprehension.

> • *Provide interpreters with a written manuscript* of your speech far in
> advance of a formal event. If you are in a consultative selling situation,
> give your prospective client adequate printed information such as bro-
> chures, diagrams, and a glossary of current industry terms about your
> product.

DELIVERY

The interpretation of body language varies widely from one country to another,
and you may be inviting disaster if you do not investigate local mores and
customs.

Although I have suggested mingling with your American audience before
and after a presentation, it behooves you to discover subtle local customs to
indicate if this is appropriate. Shaking hands is very etiquette-oriented. Even an
American child is taught to extend a hand in greeting. However, the hierarchy or
gender in many countries will determine who should initiate the handshake or
make the first verbal comment. In Asian countries, a bow may be expected. In
France, one of my clients was kissed on each cheek.

Physical proximity/distance from your listeners in small groups or one-on-
one can be interpreted as an intrusion or a rejection. I have encouraged you to
involve the audience within the first ninety seconds. A foreign audience will
resist interaction at this early stage. Wait until you are well into your presentation before posing a question or requesting *participation.*

Direct eye contact can be considered an invasion of privacy in some cultures
such as the Philippines. In general, it is better to sweep the audience with your
eyes rather than embarrassing someone by looking at them for too long.

Care about your audience. Be genuine and sincere. This appealing authen-
ticity will distinguish you in a multicultural lineup of speakers.

Voice

- *Use a normal conversational tone.* Volume will *not* make you more understandable.

- *Use variety in pitch to aid comprehension.* A monotone will add to misunderstandings in any language *if* the audience remains awake. For example, indicate contrast between two concepts by changing the pitch of your voice for emphasis, while saying "On *one* hand... on the *other* hand...." Conclude each sentence with a downward inflection instead of running into the next thought.

- *Use crisp pronunciation* by clearly articulating the endings of words. Clarity of diction will contribute to your being asked back again.

- *Slow down your pace and use pauses.* They are welcomed by an international audience.

- *Enhance your words with gestures* to underline/describe what you are saying. For example, spread your hands apart to indicate height or width. *Be careful not to point or use your thumb.* Your body language, facial expressions, tone of voice, and words should all say the same thing.

Q&A

- *Encourage questions*, but don't take the lack of response as a reflection of the value of your speech. Some audiences will be extremely reticent. Other cultures will challenge you and demand more factual data. It is important that you be available after the speech. Attendees may be embarrassed to ask questions, because they don't want to display their limited English in front of the group. However, they will ask questions one-on-one.

Some Final Suggestions

- *Contact a peer from your professional association in science or engineering who is familiar with your audience's culture* to review your presentation, including your visuals. If your company has a local office in the country, fax copies of your visuals for review. Your local university or community college may even have native-born faculty to assist you. Seek more than one opinion to get criticism from various socioeconomic perspectives.

- *Hire a translator* if you think there will be any problems with understanding your subject matter. You should speak directly to the audience and then listen to the translator. In simultaneous translation, you will speak directly to the audience while the audience listens on earphones to the translation of your words and visuals.

- *Emphasize building relationships and establishing trust.* You, as a salesperson, will be cordially received worldwide if you are able to convey useful information rather than pushing a sale.

- *Err on the side of formality.* The U.S. casual approach to business presentations can be perceived as incompetence.

- *Manage your physical environment as best you can, and then graciously reconcile yourself to your host's environment.* People can be less flexible abroad. Facilities may not always be equal to U.S. standards, but acceptable to your audience. Request everything in writing; confirm, but have Plans B and C.

- *Learn a few basic phrases or label dominant parts of a visual in the host country's language.* Such attention to detail will help bridge the chasm between you and a global audience. It will also show that you customized your visuals and didn't just recycle a talk you gave in the U.S.

Relax, work with what is offered and enjoy the differences when presenting in another country. And just think of the anecdotes you will gather for future presentations!

KEY IDEAS

- Conduct a thorough audience analysis to discover cultural values and attitudes.

- Eliminate use of confusing content, words, grammar, American slang, or idioms.

- Use visuals as much as possible to explain technical concepts.

- Include visual and verbal references to the location and interests of your audience.

- Verify that visual aid equipment is available and compatible with your accessories or be safe and take everything with you, including a low-tech version of your presentation.

Chapter 22

ANALYZING MODEL COMMUNICATORS

"Circumstances? I create my own circumstances!"
—Napoleon

OVERVIEW

Analyzing why some individuals are clearly outstanding, successful, and effective communicators will help you improve your own speaking skills. A model communicator makes choices that work, especially in difficult, stressful situations. This chapter examines some of the behavioral traits of model communicators and suggests how you can become one yourself.

Marshall Ferdinand Foch, the brilliant commander-in-chief of the Allied armies in World War I, was trying to stop a powerful German drive to capture Paris. In the heat of the second battle of the Marne in 1918, he sent a message to the Paris High Command, saying, "My center is giving way, my right is pushed back, situation...excellent. I am attacking!"

In spite of the desperate situation, regardless of what was happening all around him, Foch took control, communicated strength, and eventually led his men to victory. Although you may not face the extreme danger of a battlefield in your profession, we have all experienced stressful situations, such as a meeting with important customers on their turf, communicating with people who begrudge us their time, or making a presentation in an unfavorable environment.

I recall watching a model communicator and how he skillfully dealt with less than ideal circumstances. Robert Ballard, the president of the Institute for

Exploration in Mystic, Connecticut, was scheduled to speak at the Kingdome during the Seattle Boat Show.

After fighting a jostling crowd to gain entrance, I stopped to watch hundreds of potential buyers crawling in and out of the boats lining the floor of the stadium. Loud music was pulsating and the din of voices on the trade show floor reverberated in the huge space. I asked a yacht salesman if he could direct me to the room for Ballard's lecture on his discovery of the Titanic and the Bismarck shipwrecks. The young man motioned toward the upper levels of the stadium. "It's up there somewhere," he said. I spotted a blue drape covering a movie screen high up in the bleachers. People were grouped behind the screen facing the yachts. I worked my way toward them.

Once Ballard was introduced, he began to speak passionately about science and his challenging discoveries. He showed incredible film footage. The noise and bustle of the Kingdome faded into the background as we were drawn into his quest for these two ships.

Ballard didn't draw attention to the boat show or the other distractions of the speech site. He simply focused on why he was there and his search for the sunken vessels. In spite of an extremely difficult situation, he made his presentation a memorable experience. He made choices that worked.

HOW CAN YOU BECOME A MODEL COMMUNICATOR?

In any skill, whether it is tennis or golf, you can cut down the time needed to perfect your technique by studying the performance of masters. Model communicators are experts in their fields, and their audiences perceive them to be trustworthy. This trustworthiness and expertise combine to create credibility. In general, model communicators display the following behaviors.

- *They make an extra effort to thoroughly plan and prepare their presentations.* They use some form of the ten steps recommended in this book.

- *They start from where the audience is, not from where they are.* They begin at the same level of knowledge as their audience and find common ground.

- *They take responsibility for the audience's ability to understand the topic.* They have a gift for taking a large amount of material and breaking it down into smaller, cohesive units that can be easily understood, remembered, and applied by the audience.

- *They personalize science and technology.* Model communicators use their own experiences to annotate their materials as richly and elaborately as possible. They know that war stories and anecdotes can engage the audience, and their listeners will retain more of what they hear. A physicist told me, "Science is a personal endeavor and must be conveyed in a personal manner. You miss the resonance if you deal only with the facts."

- *They illuminate and give insight rather than dilute scientific and technical information.* "A nontechnical audience will never be at my level in my field," a scientist told me, "but that's what makes me an expert. Rather than being arrogant about my knowledge and experience, I try to give general audiences a glimpse of some of the fascinating research that's going on."

- *They have a natural style that is often like a focused conversation.* They use variety and drama in their voice and body language to capture and keep attention focused on their complex technical and scientific subjects.

- *They feel comfortable with themselves, their material, and the situation,* and feel free enough to get involved with their audience. Their self-confidence is high, and they are responsive to feedback.

Physicist Phillip Morrison of MIT (Massachusetts Institute of Technology) describes himself as "very sensitive to what is going on in the audience or their lack of attention. I often put in a local reference or a small joke to see if they are with me. I go largely by sound feedback: breathing, muffled comments, the noise of people shifting positions. Based on what I hear, I may repeat a point or drop a point." The next time you give a presentation, *listen* to your audience. What you hear will provide you with valuable clues to help you adjust your material and delivery.

Making Choices

Model communicators make choices that work. Chicago Bulls' basketball superstar Michael Jordan knows all the rules, but that's not what makes him a great all-around player. He evaluates the entire situation and makes split-second, creative choices that another basketball player might fail to consider.

People who "read" other people may be the most persuasive people in the world. Model communicators are sensitive to body language, vocal tone, and the words of others. They are flexible enough to change their behavior to get a desired response.

We can respond in various ways to difficult circumstances. What happens to your body language when you arrive at the speech site and discover no microphone, inconvenient electrical outlets for your audiovisual equipment, and fewer people than expected? Do you become defensive, angry, or resign yourself to failure? Does your voice reveal irritation, disappointment, or anxiety? Model communicators make *productive* choices when confronted with problems. Be aware of any automatic negative reactions and convert these reactions into positive responses. Acknowledge the obstacles, but promptly switch into a *problem-solving* mode. Put the situation into long-range perspective. Remember that you have choices; why not try humor?

How you act and communicate under pressure is very visible, and the rewards or negative consequences can be immediate. The ability to cope and communicate effectively under stress can make the difference between being an average presenter and a compelling one.

Having Options Available

During the San Francisco earthquake in 1989, the upper level of the double-decker Nimitz Freeway collapsed, killing more than fifty people. It was a catastrophic structural failure. Even today, in the age of computer-aided design, every part of a structure is only as safe as the weight and strength variables used in the structural analysis. Engineers know they have to provide for contingencies, and they factor these in. Nonetheless, there is an uncertainty factor within the margin of error; you can't predict every contingency.

We would like to have one set of rules that apply to all speaking engagements, but every situation is different. We can anticipate a variety of problems, but can't predict everything that might occur. Start out with a plan, but be prepared to modify it as the event transpires. Be willing to tackle the unexpected. You have choices—your choice may limit you or increase your effectiveness.

One of my clients was told that his audience would consist of staff interested in basic information about computer software. Before he began his speech, he was savvy enough to ask the audience specific questions about their background. To his surprise, he discovered that several computer maintenance people were present. He started out his presentation with general information to include everyone and was prepared to delve into some specific examples. However, he realized he was losing some people in the audience when he mentioned anything technical. On the other hand, the computer maintenance staff was clearly bored with the basics. He announced that he would present introductory information for the first part of his presentation and then, after a break, open up the forum for technical questions. Those not interested in the more specialized aspects could leave during the intermission. It was a difficult call to make, he explained, but he felt it worked out favorably.

Sometimes you will find that your audience at an association meeting will have widely divergent levels of knowledge, in spite of the fact that the published program clearly states the complexity of your subject. Engineers and scientists generally agree that you should present your paper as you prepared it, since the audience expects the level of sophistication stated in the program.

Passion and Charisma

The model communicators I interviewed had a passion for their particular field of endeavor and were motivated to tell others about it. Henry Ford said:

> You can do anything if you have enthusiasm. Enthusiasm is the spark in your eye, the swing in the gait, the grip of your hand, the irresistible surge of your will, and your energy to execute your ideas.

This fascination and involvement with their work are evident in their delivery. They renew this excitement and inquisitiveness in every presentation and follow author J. Samuel Bois' advice:

> The important thing is to live over again the experience you had when the truth that you are now expounding to the audience became a discovery, an insight, a fresh learning experience for you as an individual.

For example, Adolphus "Doc" Cheatham, a 91-year-old jazz trumpeter, was eulogized by one of his musician friends:

> For all the talk about improvising, jazz players more importantly need to be able to state a melody with clarity and passion, as if they were speaking it for the first time. Like his lifelong inspiration, Louis Armstrong, Doc could do that. Also like Louis, Doc could tell a story when he put his trumpet to his lips. As the notes flowed out, you felt they were following each other like words in a sentence. They made sense, they somehow seemed inevitable.

Many model communicators stress that they are constantly learning from each speaking experience and enjoy seeking new ways to convey complex information. They seek to develop a *relationship* with their audience. They encourage participation by getting out from behind the lectern, by walking into the audience, by calling people by name. They use the personal pronouns, "I," "we," and "you" to create intimacy. They describe their feelings and invite the audience to share their viewpoint on the emotional as well as intellectual level. And *they care about their audiences*. If you are not concerned with serving your audience, there are no words in the English language that can help you communicate. That sincere concern for others takes a deep energy commitment.

Charisma is exhibited through personal confidence, as opposed to job confidence, and a sense that you know what you are doing. Charismatic speakers always appear comfortable, no matter what the circumstances. I've seen speakers trip coming up stage steps, knock over a tumbler of water, drop notes, or have equipment fail to work. But because they were calm and kept the energy going to their audience until they recovered, these snafus were soon forgotten. Roger Ailes, communication consultant for such luminaries as President George Bush, says that *"a charismatic person never auditions."* Model communicators project confidence about themselves and their material, but they don't convey arrogance or try to compete with their audiences.

> *"Once you achieve good communication with yourself,*
> *you can communicate more freely and effectively with others."*
> —Carl Rogers

Communication is a two-way channel between the speaker and listener. But there is also another active channel: the dialogues that you have with yourself. "Do I look all right?" "Will I get through this?" "Hey, this is fun!" See the audience as a partner, and, at the same time, reassure yourself. You're the expert and you have good information. Model communicators talk to themselves in positive but realistic terms. If things aren't going well with their speech, they don't hesitate to switch styles, ask questions, or depart from their notes. Don't allow the critic or doubter in your head to sabotage your performance. Be careful about comparing yourself to the previous speaker. And even if you see someone in the audience yawning or leafing through the handouts, there is no need to become self-conscious or tense.

Balancing Work With Other Interests

People who have strong interests other than their technical and scientific professions are the ones who can call on a variety of references and comparisons to aid their imagery. Ilya Prigogine, Nobel Prizewinner in chemistry, was educated in the classics, history, philosophy, and music. Colonel Milton Hunter, of the U.S. Army Corps of Engineers, has a background in architecture and says that as a speaker, he is conscious of the structure and design of every presentation. When Henry Ford and Thomas Edison were visiting the home of Luther Burbank in the 1920's, Burbank asked his visitors to sign his guest book with their name and address. A space was marked "Interests?" and Ford watched Edison write "*Everything.*"

I spoke with an aerospace project manager who is an avid photographer, a physicist who is an opera lover, an engineer who rock-climbs all over the world, an expert in artificial intelligence who is an organist, and a programmer who grows exquisite roses. They all seemed to be able to add more variety and creativity to their communications because of their varied avocations.

Massachusetts Institute of Technology conducts more than 400 events a year in dance, music, and theater. Dr. Alan Brody, Associate Provost for the Arts at MIT, comments, "I think it is even more important to have the arts at MIT than at other places. If you focus solely on engineering, it gives you too narrow a view of life."

British philosopher C.P. Snow remarked that it is as necessary for a truly cultured person to become acquainted with the second law of thermodynamics as with the plays of Shakespeare. "The Shakespeare canon and the laws of thermodynamics are, each in their own way, among the glorious artifacts that the mind has produced," Snow stated, "and neither is more humanistic than the other, for it was humans who created them both."

James E. Olson, chairman of the board of AT&T, agrees:

> In today's world, we need the synergy that comes from an intermingling of ideas and perspectives. The engineer who's a closet poet, the administrator who ducks out to heavy metal rock concerts, the illustrator who's an inventor on the side. These are role models for the times, people soaking up the variety of life today.[1]

Explore the arts and develop hobbies that are markedly different from your profession. Everything can seem to be revolving around your intense work load, the proposal rejection, the computer crash, and the presentation coming up. Your thinking goes into a tailspin, your communications are affected, and your relationships deteriorate. But if you *dare* to take the time to remove yourself and focus on some *creative* act, you will return to work nurtured and with a broader perspective. Remember:

Good ideas are never born in the wrong places.

Observe Model Communicators

One of the best ways to learn how to be a model communicator is to study the behavior of successful communicators. Analyze what they do and how they establish rapport with an audience. It will be obvious that they have taken time to discover what motivates and inspires their audiences. How are model communicators able to communicate their ideas clearly, accurately, and concisely? Note how they make choices based on the audience and the situation. Observe the use of stories, case histories, examples, comparisons, and personal experiences.

Read good speeches. The biweekly *Vital Speeches of the Day* records entire speeches of some of the best and brightest. Analyze why the speaker chose a certain organizational pattern, why he previewed his main points, or how he incorporated humor into a serious topic. Check out Web presentations to read content and critique visuals, as well as observe delivery styles. Learn by evaluating the delivery skills of keynote speakers at association meetings or well-known communicators who visit your city to speak. Attend city council meetings, or watch press conferences to see how people respond to controversial questions or quantities of complex material. Borrow techniques from model communicators and *adapt* them to your presentation style.

SECRETS OF MODEL COMMUNICATORS

In my interviews with hundreds of model communicators, I questioned them about their secrets for success. Here's what they said:

- *Start keeping a speech file* with cartoons, jokes, quotations, anecdotes, and personal experiences.

- *Be a storyteller* and tell your story as if you were sitting at the kitchen table talking to a friend.

- Write out your speech but *write for the ear*. Carefully select words and images that call up vivid associations for your audience.

- *Deliver ideas*. An audience wants to know what you think, not exact wording from a written text.

- *Increase your vocabulary* by reading well-written prose. It will soak in and become part of your communication style.

- *Keep the adrenaline flowing.* Don't try to get rid of all your tension.

- *Avoid qualifying your viewpoint* with weak language.

- *Use active verbs.* They will demand more exciting subjects.

- *Begin emotionally on a low key* so you can build toward a dramatic conclusion.

- *Avoid* any kind of story that is even *mildly offensive.*

- *Use short, pithy quotations* that are memorable, powerful, and insightful.

- *Edit, edit, edit.* When you think you're finished, cut some more! There is no need to pass on all the information you know; no one wants to hear it.

- *Use visuals* to explain or highlight ideas, but don't let them dominate the presentation.

- *Imagine* that your presentation is a *dramatic play* and you are the scriptwriter. Consider the time, the place, the scene, what's happening before the play begins, and what happens after the curtain falls.

- *Use strong eye contact* to create the impression that you are talking privately to each *individual* in the audience instead of to a group.

- *Be efficient.* Speech preparation is hard work, so work smart. Make effective use of limited time or resources. *Follow a system.* Eliminate needless revisions.

- *Listen to an audiotape of your rehearsal.* You can make intentional changes and edit more effectively than by merely thinking through your presentation.

- *Videotape* and *critique* your presentation. It is the fastest way to improve.

- *Evaluate and learn* from past experiences. What can you do differently the next time?

Model communicators are interested in making something happen as a result of their communication. They are precise, well-ordered, and at ease with themselves. They have something of value to say and they say it in a way that deserves your attention. These are qualities that are certainly worth emulating!

KEY IDEAS

- Effective communication is never an accident.

- Have options available for the times when things go awry.

- Use your time and resources efficiently.

- Balance your professional life with outside interests and use imagery from these activities to illuminate your subject.

- Study and learn from model communicators.

Notes

1. James E. Olson, "The Spur of Ignorance," *Vital Speeches of the Day* (March 15, 1988).

Chapter 23

SHAPING THE FUTURE

"The only way to predict the future
is to have the power to shape the future."
—Eric Hoffer

OVERVIEW

Revolutionary changes in science and technology have produced equally revolutionary changes in how we think and work together. This chapter discusses the need for teams of experts from many disciplines and cultures to communicate and cooperate in order to contribute to our quality of life. It also emphasizes the strategic advantage of effective communication in any successful business. Practice the skills found in this book to help you effectively inform, sell, motivate, negotiate, market, grow, and prosper in your chosen career. Becoming an effective communicator can minimize problems when dealing with your customers, staff, and peers. It can make your work easier and more productive. It will enhance your public image. Embrace the opportunity to shape your future and the world around you!

Fifteen nations are working together to construct an international space station by 2002. The survival of the men and women who inhabit this space station will depend on their ability to integrate their diverse personalities, nationalities, and professional skills. They will need to develop a common vocabulary and learn to understand and accommodate each other's point of view. We hope this work will advance our ability to live in harmony with one another on earth.

The space station is a microcosm of our everyday world. Global communications have condensed time and space. It is critical that professionals in different disciplines in our borderless world be able to understand and work together harmoniously. Scientists, engineers, and technicians must also carry on a dialogue with the general public if they are to gain acceptance for their ideas; their new roles not only demand that the content of their presentations be of immediate value and use, but that their messages be presented in a compelling fashion. The universal stage demands effective communication skills for professional survival and success.

Scientists and engineers have been accused of communicating as if they are plugged into a headset, hearing their own personal tunes. For example, there are wide communication gaps among the scientists who live in a largely theoretical world, the engineers who design a product, and the manufacturers who produce a profitable product. An equally broad distance separates the engineers and the investors who fund the project, and a similar gap exists between these groups, the politicians, and the general public. Dan Eramian, VP of Communications for Biotechnology Industry Organization, addressed an BIO international audience in Mexico City:

> Promoters of our industry say that investment dollars and government regulatory reform are the only keys to further development of biotechnology. *They are wrong.* A third key is needed. *Effective communications* with the public, and the media and amongst the biotech community will be the most important factor in the ultimate acceptance of biotechnology on all continents. Public communications, which is not always high on the agenda of a businessman or scientist, is biotechnology's Achilles' heel.

Perceptive communicators in all technical professions should seek ways to diminish the gap between their knowledge base and that of their audiences. Innovative scientific discoveries and technological advances are not always welcomed by the public. The general population's resistance to programs promoting science and technology often occurs because questions about final proof, relative cost, and ultimate effects go unanswered. The public is frequently exposed to a flawed database of half-truths, rumors, and a clutter of statistics, and people end up bewildered, wary, or angry. Honey Rand, President of Communication Solutions, cautioned the American Hydrologists Association's International Conference:

> When the public expects to be engaged in the public policy process, the competitive edge you need, to explain your work, is communication skill.... In the absence of information, people make stuff up. In the presence of information they cannot or do not understand, people ignore it and rely on their emotions to fill the gap. Once emotions are engaged, the opportunity for dialog is frequently lost.[1]

Scientists and technologists usually stay hidden behind the scenes until a crisis occurs. Then they are called upon to explain why a bridge or building collapsed, why a species of whales is dying, or how a community beach got so polluted. Much of the defensive, hasty communication in the midst of a disaster poses more questions than answers, and has a negative long-term effect on the public's confidence in science and technology.

It is critical for these professions to continue increasing visibility and awareness of their work on a daily basis. Currently, the World Wide Web offers a remarkable opportunity to educate the public. Robert Ballard has kept school children in daily contact with his deep sea explorations through the JASON Projects. During the Mars Pathfinder expedition, NASA's Web sites received over 47 million hits in one day. High-tech companies are offering tutorials over the Web, and scientific subjects and concepts are introduced to preschoolers on CD-ROM.

Douglas E. Olesen, CEO of Battelle, a contract research institute, urged biotechnical professionals to ensure a scientifically literate society for the future, and to promote an understanding of the benefits of science and technology in the community:

> This is not simply an issue of philanthropy. It's an important business issue, critical to public acceptance of scientific advances, and it is critical to our ability to succeed in tomorrow's marketplace. We must enhance our image by balancing our time in the lab with time in the community. Our task is communication. We know we have the ability to make the world a better place; we have to make sure the rest of the world knows that too.[2]

COMMITMENT TO COMMUNICATION

> *"Prizes don't go to the people who predict rain.*
> *Prizes go to the people who build arks."*
> —Unknown

Model communicators are rare in science, engineering, and technology. Tim Bird, a model international communicator in biotech, emphasizes that, "The ability to communicate is critical to a scientist's success." Even though everyone agrees that communication skills are a number one priority, very few are willing to commit themselves to refining their talents. Some of the most important factors affecting a technical or scientific professional's career will be based on that person's ability to present ideas orally.

How many times have you been tempted to submit an abstract to the program committee of an association meeting or a technical conference? Have you ever listened to the papers being presented and thought to yourself, "I could do that. That speaker isn't more intelligent or creative than I am." As star hockey player, Wayne Gretzky, reminds us, "You miss 100 percent of the shots you never take."

Take advantage of every opportunity to present your ideas. Your confidence and ability will grow with every experience. Commit yourself to becoming a good communicator—your participation in the exchange of information will bring you satisfaction and pride. Not only will you find that improving your communication skills affects the quality of your personal and professional life, you will discover that people who are able to get up and speak about a subject are in demand. And if you can communicate well, you will be hailed as an expert, a leader.

BIGGER GAMES AND HIGHER STAKES

Several summers ago, I visited my parents in upstate New York. One night, around midnight, I was working on a speech at the kitchen table when my eighty-two-year-old dad returned from a fishing trip. He invited me to join him in a game of cards. When I told him that I had to complete a speech, he mused, "You know, there are a lot of similarities between making a speech and playing cards." I asked him to explain.

> I like the challenge, the excitement, and the uncertainty of not knowing exactly how the game will come out. It helps to be at the right place at the right time playing the right cards. Study and learn the rules of the game. Learn as much as you can about the other players. Be sensitive to their responses. Watch their eyes! Timing is crucial. Be prepared. Assess how much power you have and know when to move. Be willing to take risks and gamble. Otherwise, the game is boring. Calculate the rewards versus the risks and learn to minimize your risks. Play the hand to your fullest ability. You won't be a winner if you're afraid to lose. If you have a setback, don't give up. Bluff when necessary. Remember, there is always a certain element of luck. Commit yourself. Concentrate, and be highly motivated to win. (Then Dad smiled.) Winners get asked to go on to bigger games with higher stakes.

To prepare ourselves for bigger games, we can seek out model communicators. Their experiences and insights provide valuable lessons. Many years ago at a birthday party honoring Charlie Chaplin, the comedian entertained his guests by imitating his friends. Finally, he sang an exquisite aria from an Italian opera. "Why, Charlie, I never knew you could sing so beautifully," a guest exclaimed. "I can't sing at all," Chaplin replied. "I was only emulating Caruso."

When we use the best communicators as our models, we can bring out the best in ourselves and raise our performance to a higher standard. All you have to supply is the motivation and commitment to make your next technical presentation your best ever. Here's to *your success* in bigger games and higher stakes!

KEY IDEAS

- Commit yourself to breaking down barriers to communication.

- Build visibility and awareness of science and technology in the public eye.

- Emphasize communication skills within your organization as part of a smart business strategy.

- Take advantage of every opportunity to present your ideas and refine your speaking ability.

- Open professional doors with effective communication skills.

Notes

1. Honey Rand, "Science, Non-science and Nonsense—Communicating with the Lay Public," *Vital Speeches of the Day* (February 15, 1998).
2. Douglas E. Olesen, "Commercializing Agricultural Biotechnology," *Vital Speeches of the Day* (October 1, 1990).

SUGGESTED READING

Abrams, Marc, (ed.). *The Best of Annals of Improbable Research (AIR)*. New York: W.H. Freeman and Company, 1997.

Adams, Scott. *A Dilbert Book—Still Pumped From Using the Mouse*. Kansas City, KS: Andrews and McMeel, 1996.

Ailes, Roger. *You are the Message*. Homewood, IL: Dow Jones-Irwin, 1988.

Andrews, Robert, (ed.). *Columbia Dictionary of Quotations*. New York: Columbia University Press, 1993.

Applewhite, Ashton, William R. Evans III, and Andrew Frothingham. *And I Quote—The Essential Public-Speaking Resource*. New York: St. Martin's Press, 1992.

Axtell, Roger E. *Do's and Taboos of Hosting International Visitors*. New York: John Wiley & Sons, 1990.

Beckwith, Harry. *Selling the Invisible*. New York: Warner Books, 1997.

Boettinger, Henry M. *Moving Mountains (or the Art of Letting Others See Things Your Way)*. New York: Collier Books (MacMillan), 1969.

Braganti, Nancy, and Elizabeth Devine. *European Customs and Manners*. New York: Simon & Schuster, 1992.

D'Arcy, Jan. *Speak Without Fear—How to Give a Speech Like a Pro*. Chicago: Nightingale Conant, 1987. (Audiocassettes).

_____. *Dr. Jack's Adventure in Videoconferencing Land—A Guide to Communicating Effectively on Camera*. Bellevue, WA: Jan D'Arcy & Associates, 1990.

Dayton, Doug. *Selling Microsoft—Sales Secrets from Inside the World's Most Successful Company*. Holbrook, MA: Adams Media Corporation, 1997.

Fisher, Roger, William Ury, and Bruce Patton. *Getting to Yes,* 2nd ed. New York: Penguin Books, 1991.

Humes, James. *Podium Humor—A Raconteur's Treasury of Witty and Humorous Stories*. New York: Harper & Row, 1975.

Hyatt, Carole. *The Woman's New Selling Game—How to Sell Yourself and Anything Else*. New York: McGraw-Hill, 1998.

Leech, Thomas. *How to Prepare, Stage & Deliver Winning Presentations*. New York: AMACOM, 1993.

Leong, Ming, Swee HongAng, and Chin Tiongtan. *Insights for the Asia Pacific*. New York: Heinemann, 1996.

Morrisey, George L., Thomas Sechrest, and Wendy B. Warman. *Loud and Clear—How to Prepare and Deliver Effective Business and Technical Presentations,* 4th ed. Reading, MA: Addison-Wesley, 1997.

Morrison, Terri, Wayne A. Conaway, and George A. Borden. *Kiss, Bow, or Shake Hands—How to Do Business in Sixty Countries*. Holbrook, MA: Bob Adams, Inc., 1994.

Morrison, Terri, Wayne A. Conaway, and Joseph J. Douress. *Dun and Bradstreet's Guide to Doing Business Around the World*. Paramus, NJ: Prentice Hall, 1997.

Noonan, Peggy. *Simply Speaking—How to Communicate Your Ideas with Style, Substance and Clarity*. New York: Regan Books, 1998.

Presentations—Technology and Techniques for Effective Communication. 50 S. Ninth St., Minneapolis, MN 55402. Monthly.

Seidler, Ann, and Doris Blain Bianchi. *Voice and Diction Fitness*. New York: Harper & Row, 1988.

Tannen, Deborah. *You Just Don't Understand—Women and Men in Conversation*. New York: William Morrow and Company, Inc., 1990.

Technical Communication. Washington, DC: Society for Technical Communication. Quarterly.

Tufte, Edward R. *The Visual Display of Quantitative Information*. Cheshire, CT: Graphics Press, 1983.

_____. *Visual Explanations—Images and Quantities, Evidence, and Narrative,* Cheshire, CT: Graphics Press, 1997.

Wilder, Claudyne, and David Fine. *Point, Click and Wow!!—Quick Guide to Brilliant Laptop Presentations*. San Diego, CA: Pfeiffer & Co., 1996.

Woelfle, Robert M., (ed.). *New Guide for Better Technical Presentations*. New York: IEEE Press, 1992.

INDEX